American Sunshine

American Sunshine

Diseases of Darkness and the Quest for Natural Light

DANIEL FREUND

The University of Chicago Press

Chicago and London

Daniel Freund is assistant professor of social sciences at Bard High School Early College.

The University of Chicago Press, Chicago 60637
The University of Chicago Press, Ltd., London
© 2012 by The University of Chicago
All rights reserved. Published 2012.
Printed in the United States of America

21 20 19 18 17 16 15 14 13 12   1 2 3 4 5

ISBN-13: 978-0-226-26281-9 (cloth)
ISBN-10: 0-226-26281-2 (cloth)

Library of Congress Cataloging-in-Publication Data

Freund, Daniel.
     American sunshine : diseases of darkness and the quest for natural light / Daniel
Freund.
          p.   cm.
     Includes index.
     ISBN-13: 978-0-226-26281-9 (hardcover : alkaline paper)
     ISBN-10: 0-226-26281-2 (hardcover : alkaline paper)
     1. Sunshine—Environmental aspects. 2. Urban ecology (Sociology)—United States.
3. Climatotherapy. I. Title.
QC911.F83 2012
577.5′6—dc23

                                                                                    2011034927

♾This paper meets the requirements of ANSI/NISO Z39.48-1992 (Permanence of Paper).

CONTENTS

ACKNOWLEDGMENTS

Though any list of thanks will be incomplete, I would like to recognize a few of the people and institutions responsible for helping make this book possible.

Fellowships from the Lemelson Center for the Study of Innovation and Invention, the Smithsonian Institution, and the Herbert H. Lehman Center for American History were essential to the completion of this project. The first two of those awards also gave me access to Washington's considerable resources. DC institutions house remarkable materials, essential to my treatment of this subject; these places also house live minds, who helped me find the materials there to be found and think in ways I had not thought to think. Among these many great people, Susan Strange, an archivist at the National Museum of American History, deserves special recognition. Her knowledge of the Smithsonian's diverse collections and eagerness to make connections between my work and its resources improved this book immeasurably.

Betsy Blackmar and Kenneth T. Jackson provided just the assistance I needed to shape this work: constructive criticism and consistent encouragement. Without them, this book might still have happened, but it would have been far less in their absence. Their suggestions improved my thinking and helped me become a better historian. Thanks also to Christian Warren and James Colgrove, who offered smart comments. My editor at the University of Chicago Press, Robert Devens, is largely responsible for this publication. He approached me when I was stuck, offered the criticism necessary to move me forward, and sent my work to the two reviewers—whoever they are—whose suggestions pushed me in new directions. He also introduced me to Anne Goldberg at the Press, whose patience with my unending questions was remarkable.

There were a few people close to me who read this book and gave me their feedback. Nancy Kwak offered well placed suggestions and much needed positive reinforcement. To my mother, my father, and my wife, Meg—I believe you when you say you enjoyed reading this work, but we all know that you probably would not have picked it up had I not been its author. I owe you thanks for so much more than comments on my prose: you have loved me unflaggingly and provided the support necessary to help me finish this project; you have made me smarter, and you have made me better. And finally, Baby Eli, one day you may read this book. If you do, please know that you have made the last year plus the most joyous of my life. Know also that your good temper allowed me to finish this manuscript and that the prospect of time together gave me the energy to rise before the sun.

Toward a History of Natural Light

Around 2,500 BC, the Egyptians elevated Re, their sun god, to a hallowed place in their pantheon. He became a creator and father of pharaohs. More than a thousand years later, Akhenaten decided that Aten, another solar deity, should be even more important. The pharaoh proclaimed that Aten was the one true god and thus, according to some, gave birth to the civilized world's first monotheistic religion.

A few decades before the birth of Jesus, Vitruvius wrote *Ten Books on Architecture* for both a technical and lay audience. He told of the proper geometry for an atrium, the center of the Roman home, open to the sky and light, which would help illuminate the entire house. Smart builders, according to the architect, oriented winter dining rooms for better early evening light, pointed libraries east so that they were bright in the mornings, and located windows to avoid the shadows of neighboring buildings.[1]

In 1758, Carol Linnaeus, probably Sweden's greatest scientist, completed the tenth edition of his *Systema Naturae*. The original version, which offered the classificatory system for plants and animals still in use today, is among biology's seminal texts. In edition ten, Linnaeus provided two names for humans, *Homo sapiens*, thinking man, and *Homo diurnus*, day man. Though he did not elaborate on the second characterization, the Swedish scientist contrasted it with humanity's closest relative, *Homo nocturnus*. This species identified the troglodytes, cave dwellers, "children of darkness, who turn day into night and night into day . . . not much larger than a boy of nine years old; white in color, and not sun-burnt, because they always go about by night."[2]

These three periods, separated by gulfs of time and topic, dramatize the universality of sunlight as a subject for consideration. This book introduces modern American analogues to Akhenaten, Vitruvius, and Linnaeus—figures who worshipped the sun, tried to redesign the home for brightness, and advocated a nature for man in which daylight figured prominently. There is no reason, however, to assume a direct cultural lineage from ancient Egyptians—or for that matter, from Celtic Druids, Mesoamerican Aztecs, ancient Indians, or native North Americans. A billion years older than Earth, the sun, unsurprisingly, has been a part of all cultures.

The existence of cultural responses to sunlight may be a relative constant, but the content of those attitudes has changed dramatically—and never more so than in America beginning in the middle of the nineteenth century, when cities grew darker. First hulking tenements then towering skyscrapers cast huge shadows, pollution grew dense enough to obscure the heavens, and work chased more and more people, day in and day out, into manmade caves. Suddenly, it seemed, the sun's light had begun to fade.

With these disturbing conditions growing worse, Americans complained that their dark cities bred disease, cultivated immorality, and made once-profitable spaces unrentable. The initial responses to such concerns fit well with the reform mood of the time. There were bold social solutions: urban planning and tenement and school reform, all efforts to return light to a suffering public. New Yorkers, residents of the nation's densest and darkest place, did much of this early work, but they were not alone. Others, some in metropolises and many in more modest cities, took action, fearing that darkness would spread to their homes in Milwaukee or El Paso.

Primarily focused on the city in its first chapter, this sunlight history grows outward in its second. During the 1920s, scientists determined that the solar spectrum—in contrast to most artificial alternatives—was a uniquely potent light that built strong bodies. This discovery attended one that was quite worrisome: the healthful composite of waves could not effectively penetrate glass or smog, meaning that cities, where people worked indoors and pollution darkened skies, were more hazardous than previously imagined. These later authorities ignored the evidence that sunlight's primary benefit was in the production of a *single* vitamin, D, which prevented a debilitating *childhood* disease, rickets. Adult Americans were worried about sunlight, and doctors and scientists provided compelling new hypotheses that validated those concerns. If sunlight could treat one particular condition, doctors began to wonder, perhaps it could do much more.

Deepened and refocused, concerns about the light needs of Americans grew. Industrial cities may have been at the root of the problem, but scientists soon reached the consensus that innovation would offer new promise in a frightening time. It would prevent a dark future of pasty children, stunted in their growth, and pallid adults, susceptible to illness. Lamps (not conventional incandescent lamps, which emitted none of the critical healing rays, but special sunshine replicators) or glass (not regular old glass, which shut out much of the solar spectrum, but innovative alternatives) could provide relief to the sun starved. If the old solutions to the darkness problem were Progressive and social, the new ones were technical and personal.

No place became more closely associated with the new sunlight movement than the sanitarium, but lots of other institutions also found a place for heliotherapy. Medical experts, who had long asserted a role for environmental cures in tuberculosis treatment, now claimed to have found the key to a post-tubercular world. Doctors and scientists proclaimed that sun treatments taking place in homes, hospitals, workplaces, and schools required their expert care because light was not everywhere the same, and bodies were no less variable in how they reacted to exposure. Zoos and farms, under the expert control of handlers and husbandmen, began treating their own sickly patients. These institutions and professionals, subjects of chapter 3, did not offer a singular vision. Authorities quickly learned that they could not direct a sun-starved public or control money-hungry businessmen. The public often did what it wanted when it consumed—or refused to consume—lamp and glass innovations, headed to the beach seeking some rays, exchanged modest swim suits for revealing ones, or began proclaiming a commitment to sunshine as an unparalleled healer.

The news that sunlight was essential to welfare brought all sorts of Americans together in a quest for health. Chapter 4 tells their stories. Designers again looked for ways to improve residences, government offered incentives to build for light, settlement-house workers worried that immigrants were ignorant about the light needs of their children, and eugenists decried the health of a white race hidden from the nurturing effects of sunlight. These sun seekers found themselves united in their concerns with an advanced guard of clothing experts who contended that the time had come to swap long dresses and high collars for short skirts and open necklines. Housewives asked how to care for their children, government bureaus encouraged little strapped jumpers for toddlers, department stores offered thinly woven clothing, and a small but growing nudist movement joined its concerns about modernity to the emerging evidence that big cities, office jobs, and wool overcoats were hazards to a healthful life.

There was a note of panic in much of this dialogue, but the long-term solution, featured in chapter 5, was near at hand. By the middle of the 1930s, most people had concluded that time in the sun brought health and happiness, and the tan was a marker of vitality. Florida and California boosters capitalized, claiming that health seekers should head west and south in search of climates suited to an outdoor lifestyle. For those who could not travel so far, there was a decidedly indoor alternative. By 1935, milk companies, promising all the health of a day in the sun, had begun to fortify their products with vitamin D. Even they admitted that a life lounging in the bright sun was ideal, but that was little more than a dream for most, and there was no way to store up enough light for a long winter during a short summer vacation. Milk was the ever-available, economical sun healer. Matters of class obviously play an important role throughout this history, but never more than in this twofold solution to sunlight problems. Progressive plans for universal access to sunlight had faded, replaced by alternatives to natural light accessible to those who could afford them. In the end, vitamin D milk brought some of the health benefits of sunlight to many Americans, but there remained a belief that it was a limited alternative to better, costlier solutions.

This book, which deals with a topic as basic as sunlight, necessarily sprawls. It ranges from Progressive reform to the histories of the city, the environment, medicine, science, nudism, vacationing, and parenting. It also breaks through geographic boundaries. By 1930, communities throughout the nation worried about sunshine, and plenty of southern and western locales had become willing to share their abundant natural light. America, however, was not unique: Londoners expressed worries about smog, Berliners celebrated their love of sunshine, and the new world looked to the old for models in an attempt to understand the medical wonders of light. The American response to sunlight can only be understood within the context of this international movement.[3]

It is no easier to quantify than to bound sunlight enthusiasm. This book explores practices that did not take off, as well as those that did, because the former are as instructive as the latter. Only a small percentage of Americans became nudists, and sunlamps never captured much of the lighting market. Nearly every sanitarium had a heliotherapy ward, but most Americans avoided stays in them. California became a popular destination, but then again, not everyone who went to Los Angeles sought sunlight. There were other developments with which most Americans were (and are) familiar: the tan became beautiful and healthful, the beach emerged as one of America's iconic summer-vacation spots, and dairymen nationwide began fortifying their milk with vitamin D. In the history of sunlight, the sheer diversity of topics is telling. A picture of widespread concern about the need for natural light emerges among a variety of Americans. Moreover, examination of even short-lived medical treatments and rare, reengineered lightbulbs facilitates important reconsiderations of commodities, nature, technology, and health.

At this moment sunlight developed into a scarce and valuable resource in America. Of course, there was no less light from the sun, but there *was* less for some Americans: those working deep inside big buildings or living in the shadow of neighbors. People have a history of transforming nature to make it profitable, turning prairie into pasture or plowing dense sod into productive farmland. Sunlight, however, is an unconventional commodity. Cities, where the goal was buildings not shade, made natural light valuable quite by accident. In this respect, sunlight is more like timber than farmland. Settlers generally cut down trees to build houses, plant fields, generate warmth, or erect towns. The scarcity of wood and its resulting change in value was an unfortunate byproduct of these activities. In other senses, sunlight operates more like an idea or luxury. Americans uncovered and built its value in agreements between apartment renters and their landlords, between electric companies and consumers, between Florida visitors and boosters, and in laboratories, doctors' offices, and popular culture.[4]

Progressive planners believed that there was a problem with the way people thought about this suddenly precious thing. Sunlight, they said, should not be a luxury or even a resource: it was really a necessity—perhaps even a right—little different from fresh air or clean water. They proffered a remedy but failed to solve the problem. In time, businessmen stepped to the fore, seeking profits and claiming they could recreate natural light. Whether their new products were adequate substitutes for the original was debatable—a debate scientists, doctors, government bureaus, and the public vigorously entered—but for businessmen and many consumers, innovation could return sunshine.

In considering what to do about their darkness problem, Americans developed complex senses of the relationship between innovation and nature. Historians have tried to sort out whether sentiments around this time embraced or feared modernity and how Americans sought to transform their environment while looking to make those changes appear natural (to others and to themselves). Some of this scholarship speaks of the nineteenth-century American pursuit of a middle landscape, "nature improved by a modest intrusion of man and the machine." If the middle landscape dominated much of nineteenth-century thinking about nature, by the 1930s, a different attitude was becoming ascendant. The goal was no longer a modest intrusion, but a more perfect replacement—the creation of manmade, engineered landscapes that synthesized all that was best about nature without its limitations.[5] Advocates of this vision argued that sunlight was indeed a great thing, but clouds could block it out, short days could make it scarce, and pollution could occlude health-giving rays. Nature was fickle, but artificial sunlight did not have to be. Sunlamps could be the perfect substitute for imperfect nature and so, for that matter, could vitamin D–fortified milk. The machine's intrusion had become total, but the goal was perfected "nature."[6]

From the time when American sunlight fears emerged to the moment that vitamin D milk hit the shelves, doctors, scientists, and public health authorities determined that they would definitively and accurately parse all that was healthy about sunlight, and urban planners contended that they could craft a vitalizing environment. They were often wrong in their pronouncements, and the darkness problem was frequently more imagined than real. Experts routinely misassessed risks, misunderstood their subjects, and misguided a public worried about something they perceived as a danger. While historical inquiry may render these errors understandable, they were, nevertheless, errors.

The history that follows will travel America, into slum alleys, medical wards, and Hollywood movies. It examines miracle lamps designed to recreate the light of tranquil meadows and renders comprehensible the begoggled chimpanzees lying under them. It may seem like an odd story, as at times it does to me, but such phenomena were not curiosities to be gawked at, and they are not peripheral to the twentieth century. They reflect and expose dreams of riches and hopes for a benevolent, synthesized nature. For the businessman, sunlight became a moneymaker, for the doctor it developed into a healer, for the beachgoer it emerged as a vitalizer, for the historian it can be a revealer, but first, for the tenement reformer, it was the great cleanser.

1

The Darkening City, 1850-1920

John Griscom was worried. In 1845, the sanitarian conducted the first formal survey of New York's housing conditions, and he did not like what he saw—or rather, did not like that he could not see. Bad housing meant corrupt morals, weak bodies, and dependency, burdening government and sapping the nation's vitality. Griscom found worse than bad housing. He told of residents who suffered dearly in "dark hole[s]—devoid of windows, which made fresh air and sunlight 'entire strangers to [their] walls.'" Other tenants had it worse; cellar dwellers endured near pitch-blackness. Stepping down, Griscom wrote, one "must grope in the dark, or hesitate until your eye becomes accustomed to the gloomy place, to enable you to find your way through the entry." Fumbling through "dark, damp recess[es]," only auditory cues, a flickering lamp, or a dirt-coated window could lead a visitor to tenants.[1]

Griscom's report began a crusade that would grow powerful in the coming years. Throughout the second half of his century, the city he hoped to reform would attempt to take on its problem. Generally, measures did little. In 1901, however, Albany legislators passed a tenement house law to resolve New York City's housing problem. Its goals ranged far: reduce fires, check prostitution, and decongest the slum. Its most extensive section, however, dealt with none of those concerns; it took aim at poor ventilation and dark rooms, looking to fix a city where overbuilt blocks prevented light from entering residences.

By the time Albany took its most substantive action, concern had ranged outside of New York and beyond slums. Throughout the country, in subsequent years, municipalities followed New York's tenement-reform model. Chicago was one of those places, and like Gotham, its darkness problem was not limited to slums. By the 1890s, skyscrapers had made canyons of its downtown. While the Second City was first to take legislative action against its skyscrapers, soon New York again produced the nation's most impressive sunlight legislation.[2]

Between 1850 and 1920 reformers often combined their concerns about light and air into a general condemnation of the urban environment. Nevertheless, the perception that darkness was a problem in the maintenance of morals, health, and property values was growing. With the light-is-good / dark-is-bad binary approaching its full form, reformers took action. Their solutions to disturbing conditions were hopeful, social, and organizational—in a word, they were Progressive: redesign the tenement, reorganize the city, and remake the classroom.[3] Concerned citizens thought it was irrelevant that many of the places they sought to fix were more big towns than burgeoning metropolises; darkness had spread far and the future looked increasingly bleak with industrial cities proliferating. Worried that sunlight was everywhere retreating, they decided that there was no time like the present to take action.

Griscom's report expressed profound concerns, but before long conditions were worse, and his big, dingy, scary tenements seemed almost quaint. According to an 1865 report by the Association for Improving the Conditions of the Poor, new monstrosities stood tall and packed families into rooms that were "small, dirty, very badly ventilated, poorly lighted, and wretched in the fullest sense of the word."[4] In the following decades, waves of reform came—in 1869, 1879, 1884, and 1895. Often legislation indicated big hopes and contained bold prescriptions for a brighter future. All failed. By 1900, many of Manhattan's new tenements took up nearly all of their 25-by-100-foot lots and stood six stories high. Oversized buildings on overstuffed lots left little room for open space; with structures shading each other, windows offered minimal benefits.

Many of these tenements owed their shape to an 1879 law. The light courts it required led builders to design structures like dumbbells, wider in the front and back and slightly narrower in between. With previous buildings often reaching from lot line to lot line, these new structures were an improvement, and initially public advocates hailed the innovation as a model worth emulating: a narrow side yard was better than no yard at all. Before long, however, that improvement—real though it was—became infamous, and many of the same tenement reformers who celebrated the dumbbell design spoke out against it. The problem, they contended, was mathematical. In the front and back, with exposures onto the street or a rear yard, light was rarely a problem. However, profiteering builders reduced side yards to a bare minimum, and messy tenants did not help matters. As a result, the dumbbell's handle, which included several distinct spaces—halls, stairs, bedrooms—opened onto five-foot-wide rubbish pits and almost never received light. Reformers complained that the 25-foot lot, with accommodations for four families per floor, effectively precluded light. Prominent architect Ernest Flagg overstated the case when he expressed the common sentiment that the design was "the greatest evil which has ever befell New York City," a "disaster" beyond compare that made healthful, well-planned buildings with sufficient light and air impossible. That a city of apartments built without the 1879 law would have been worse was hardly worth mentioning for turn-of-the-century reformers.[5]

By 1900, New York had swelled to almost 3.5 million people, two-thirds of whom lived in tenements. The city's housing, according to the definitive survey of the time, contained 350,000 dark rooms, a number growing as new dumbbells replaced old structures. Some blocks on the Lower East Side were among the densest in the world. One housed 2,781 people in two acres. Of its 1,588 rooms, almost 30 percent were dark, and 40 percent more only had exposures onto gloomy, narrow airshafts. More than two thousand of Manhattan's tenements were in the rear of a lot containing another building. They were the worst. Lawrence Veiller, New York's most indefatigable housing reformer, investigated a sample of the most troubling rear tenements and found that 41 percent of stairs and rooms were pitch black, 38 percent were very dark, and 21 percent were dark—a total of 100 percent.[6]

Similar conditions could be found elsewhere on the island. In 1894, the city's Architectural League president, George B. Post, said that tall buildings were an evil and that a street lined with them was "like a bottom of a canyon, dark, gloomy, and damp." But Gotham just got bigger and so did the problem. G. W. Tuttle and Herbert S. Swan's twice-published *Planning Sunlight Cities*, computed that at noon on December 21, the Woolworth Building cast a 1,635-foot shadow, and the Equitable, a recent, imposing addition to the city's skyline and a lightening rod for criticism, shaded 7.59 acres.[7]

In 1913, New York's Board of Estimate and Apportionment authorized a study to help inform a new law that would "arrest the seriously increasing evil of the shutting off of light and air from other buildings and from the public streets." The *Report of the Heights of Buildings Commission* found that the gravest problems were limited to the southern tip of the island, and only 1 percent of buildings reached taller than ten stories. Still, the document told not of a limited problem but of a considerable and expanding one. A subsequent report even provided a graphic representation in order to show what happened on Exchange Place between Broad Street and Broadway when a building cast all-day shadows. Neighbors, the caption explained, had little choice but to resort to artificial light even on sunny days.[8]

Manhattan was not alone with its concerns. A year before the *Times* spoke with Post, *Harper's* published Henry Fuller's serialized novel *The Cliff Dwellers*, which told of the social machinations in the Clifton, an eighteen-story Chicago office building. Fuller's city was a corrupted landscape of pseudo-shrub telegraph poles and mock-tree chimneys. Its air was not oxygen and nitrogen but soot: "The medium of sight, sound, light and life becomes largely carbonaceous, and the remoter peaks of this mighty yet unprepossessing landscape loom up grandly, but vaguely, through swathing mists of coal-smoke." Its buildings were "towering cliffs" with "soaring walls of brick and limestone" that jutted up along canyon-like streets. Many of the Clifton's renters occupied a world of shadow, with the bottom floors lacking sunlight except for a short while each day early in summer.[9]

While Fuller paid considerable attention to the sootiness of his environment, pollution was somewhat less of a concern than cliffs and canyons in this early period of sunshine enthusiasm. Blackened skies had begun capturing reformers' attention, but many still saw pollution as an indicator of progress, evidence that the nation was an industrial powerhouse. Some even ascribed curative properties to smoke. Those who did see cause for concern often struggled to define their issue, unsure whether they were dealing with a health or an aesthetic crisis. At this point, shade was the greater worry. As was the case in smoke-abatement arguments, it was not always clear at first just why darkness was such a grave problem. For some, lightness was little more than a vague metaphor; for others, sunshine was growing to prominence as a critical element for the prevention of disease, promotion of virtue, and protection of property.[10]

Books like *Lyrics of Sunshine and Shadow* and *The Story of My Life, or the Sun-shine and Shadow of Seventy Years* simply told stories or offered poetry about sometimes difficult, sometimes good days. *Sunshine and Shadow of Slave Life* was similar, and its author's explanation of the memoir's contents indicates his in-substantial sense of light and dark; the two mapped simply to "the best features as well as the worst" of his life. Elsewhere, the pair's meaning was a bit more complex but still mostly metaphorical. In the 1868 book *Sunshine and Shadow in New York*, dubious characters skulk in shadow; poor people stew in filth and gloom; and good citizens, the sunshine, take action. The text's author, Matthew Smith Hale, exuberantly celebrated work that joined religious with social causes because, as he put it, "great cities must never be centres of light and darkness; repositories of piety and wickedness; the home of the best and the worst of our race; holding within themselves the highest talent for good and evil, with the vast enginery for elevation and degradation." Historians have rightfully cri-tiqued this simple urban portrait as fantasy, but to Hale, it seemed real. The city was a place where sunshine battled shadow for the souls of residents.[11]

For others, however, light and dark were far more than literary devices. Pro-gressive reformers worried that dark places became centers for human debase-ment, where children could not grow into good people and immigrant parents, isolated from the benevolent care of Americanizing natives, became bad citizens who kept unhygienic homes. The Progressive story was simplistic and largely inaccurate, revealing ethnocentric versions of good, clean, and American; in it, an environment of corruption and degradation festered, in part because bad be-havior thrived when hidden from light.

Many Progressives argued that darkness did more than simply hide transgres-sions; it was positively toxic. Each "crowded, ill-ventilated tenement" crippled children and bred delinquency. Only human life was cheap, with "light, air, water, heat, the elemental things cost[ing] blood money." Robert DeForest, chair of the committee that effectively made the 1901 housing law, claimed that trouble was built into the very definition of tenements. He quoted the ap-parently authoritative *National Cyclopedia*, characterizing them as "commonly speaking . . . the poorest class of apartment houses," too dark, poorly ventilated, and overcrowded, with many rooms that were often completely without day-light. In them, "bad air, want of sunlight and filthy surroundings work the physi-cal ruin of wretched tenants, while their mental and moral condition is equally lowered."[12]

While many, like DeForest, lumped darkness with a litany of environmen-tal toxins, others singled out its debasing effects for special consideration. An

1894 tenement house report claimed that, on the subject of morals, the only issues that interviewees consistently brought to the authors' attention were overcrowding and darkness. In the case of filth and germs, the main problem was darkness alone. A letter printed in *Charities*, a leading reform periodical, made it clear that goodness and brightness were linked: "Darkness fosters in young children all sorts of immorality, and it tends to make the persons living in such houses oblivious to filth." When light hit the hall, tenants quickly became ashamed and cleansed their buildings and bodies. In its absence, filth and moral corruption festered, hidden from watchful eyes.[13]

Nobody more effectively dramatized troubling tenements and the needs of residents than Jacob Riis. His photographs of alleys convey messages in light and shadow. The street level in his images is generally cramped and dark, almost claustrophobic, and often features immigrant toughs or children playing with little supervision. The sky, by contrast, is invariably bright, almost glowing. In indoor scenes of cramped quarters and dark rooms, of working mothers and children, Riis's work suggests the hidden moral hazards of tenement canyons. One photograph shows a child standing alone in a hallway; the caption reads, "Baby in slum tenement, dark stairs—its playground." The girl leans against a wooden board, which has pealed away from the wall. She wears a dirty dress and stands on a filthy floor. Her slightly blurred head and the rope tied around a banister give the feeling of a soul lost in chaotic isolation. The picture powerfully expresses the Progressive message: darkness means chaos, disarray, and lost childhood.[14]

However, there is a challenge to reading this photograph. Regardless of what the caption says, the photograph is not dark, and the girl is not hard to make out. Photographs distort light. They can introduce it where it does not exist—by using a flash—or they capture more of it than would the eye. The inescapable fact that photographs distort presents historians with a challenge, but not an insurmountable one. Riis's pictures may have misrepresented their scenes and most certainly were often staged, but they were quite clearly emblematic of their time, demonstrating a perspective that located danger—moral and physical—in darkness. In fact, the lightness of the scene, when paired with its caption, only adds to a sense that Riis has brought something terrible to light.[15]

Afraid of what happened hidden from view deep in slums, reformers focused closely on light, but they saw benefits well beyond the removal of moral contagions and the cleansing of cluttered hallways. Brightness brought wellness too. Griscom and the Association for Improving the Conditions of the Poor had a vague sense that the sun could help prevent tenement residents from falling ill.

Figure 1.1. "Baby in slum tenement, dark stairs—its playground," ca. 1890. Jacob A. Riis Collection, Museum of the City of New York.

Other light enthusiasts pushed sunshine as a medical wonder more aggressively. Around the same time that Griscom issued his report, but an ocean away, Robley Dunglison lamented industrial conditions in Britain that threatened young workers, contrasting the "pale deformed being" brought up in confined cities to the "ruddy native of Country Situations." In the half-century that followed, theoreticians would echo similar concerns. American Edward B. Foote Jr., a self-avowed light expert and author of *The Blue Glass Cure: How and When it Originated*, asserted that the sun was a spectacular healer, favored by the ancients and returning to prominence:

> The sun has not for a long time been accorded that homage which is due to it as the source of all our blessings; and its benign influences have been for centuries willfully and ignorantly neglected; but of late science has been prying into the origin of things, and it has traced back to the sun the source of all the vegetable and animal life which inhabits the planet.

Foote contended that America was not the only nation enthusiastic about light and singled out Englishman Forbes Winslow, author of *Light: Its Influence on Life and Health*, for special consideration. In his celebration of sunlight, Winslow bemoaned the physical and mental health of miners trapped in darkness throughout the daylight hours and made comparisons between city children and country children that would have impressed Dunglison.[16]

Despite these points of agreement, the early light discourse was fractious; most glaring, two self-assured camps battled over whether clear or tinted sunlight brought health. Foote included as part of his eighty-word subtitle, *Gen. Pleasonton Not a Success as an Experimental Philosopher*. Much to Foote's dismay, no doubt, bad science did not prevent General A. J. Pleasonton's popularity; his blue-light theory had captured a considerable following. Pleasonton argued that color had profound biological effects and that blue light contained powerful electromagnetic forces that could return bodies to health and build strong constitutions. Other theoreticians asserted biological roles for reds and yellows too, claiming that different colors had uses in curing different ailments: red stimulated the body to action, and blue calmed it. For one of these theorists, Edwin D. Babbitt, sunlight, an amalgam of all colors, contained the best of all worlds, energizing or relaxing as needed. Really, though, Babbitt was a color enthusiast as much as a light enthusiast. Blue light had effects similar to the flower foxglove and red worked much like cayenne pepper. Both he and Pleasonton struggled to understand light's effects in producing health, and neither thought that sunshine had unique properties; for one the issue was really energy and for the other it was color. Without a compelling mechanism by which to explain color's effects, Pleasonton and likeminded theoreticians struggled against a doubting—sometimes mocking—medical establishment. Their success was short lived.[17]

Before long, biologists asserted a new role for sunlight in preserving health, which diverged considerably from the claims of Pleasonton, Foote, and Dunglison. In the 1870s and before, the prevailing theory of disease held that gases, miasmas—products of unsanitary conditions—caused disease. About thirty-five years after Griscom's report, German Robert Koch, an adept experimental scientist and a clever theoretician, revised that thinking. His work—largely built on the efforts of other scientists like Louis Pasteur—isolated bacteria and enabled him to formulate four simple criteria necessary to show that a particular infecting agent caused a particular disease. It is hard to overstate the importance of the bacteriological revolution and easy to overstate its short-term transformative effects. Much of the old miasma theory endured. The public and scientists accepted contagions even as they remained confident that filth and squalor bred sickness. Miasmas did not explain microbial infection, but that does not mean people abandoned old ideas entirely.[18]

The faith that environment and in particular sunlight could be a great curative found new support in early bacteriological thinking. In 1877, Englishmen Arthur Downes and Thomas Blunt reported success killing microbes with sunlight. They also contended that not all parts of the solar spectrum were similarly effective antibacterial agents—blue was best, red worst.[19] But unlike Babbitt's research, this work made color a side note; sunshine, writ large, saved. While doctors disputed some parts of Downes and Blunt's findings, sunlight's role in killing infection quickly became doctrine. By the turn of the century, Koch was on board and bacteriology textbooks, including ones published in America, told their readers that sunlight was a critical part of medicine's weaponry.[20]

Tenement reformers picked up on this evidence and used a new language in expressing compelling concerns about slum conditions. Small side courts were giant "culture tubes," and DeForest reminded readers of scientific findings that tuberculosis thrived in dark homes. A primer on tenement inspection agreed, pointing its audience to a British sanitarian whose book featured sunlight as one of the critical means for promoting health. Downes and Koch were described as heroes who had helped to prove that "the healthiness of the dwelling in this country increases in proportion to the amount of daylight and sunlight admitted." Elsewhere, Dr. Hermann Biggs reiterated that sunlight could kill the tuberculosis bacillus, "the only necessary factor in the production of [the disease]," and Dr. Cyrus Edson, medical examiner at New York's Board of Health, told a reporter that sunlight, all agreed, was "necessary as a destroyer of germs."[21]

Arthur R. Guerard, however, saw a larger role for sunlight in the fight against disease. His article in *The Tenement House Problem*, the most impressive single work in the reform movement, claimed that an improved urban environment would prevent 8,000 to 9,000 tuberculosis deaths. For Guerard, sunlight did indeed kill bacteria, but that was not the extent of the health measure's value; it also made the body stronger. Guerard, like others before him, used the sickly, pale, urban child as evidence that bodies raised in darkness were far more susceptible to infection. He claimed that work employing sunlight and other medicine to fortify patients was underway in a great many sanitaria. Blue-light advocates might have embraced such a conclusion, but their time had passed, and other constitutional medicine enthusiasts had not yet become popular. In tenement reform and at the turn of the century, Guerard's claims about the role of sunlight as a constitutional aid were far from the norm.[22]

Though their concerns were often not health related, skyscraper critics were quick to parrot the charges of tenement reformers. A few of them even followed Guerard's thinking. For William Atkinson, a fellow of the Boston Society of Architects, sunshine, unlike artificial light, was a highly potent therapeutic agent, with evidence of its efficacy long settled by German and Danish investigators. It killed disease and enabled the body to fight infection. The *Final Report* of the Commission on Building Districts and Restrictions made the argument even more forcefully: the relationship of light and air to the public's health was "immediate and undoubted." Sunlight and fresh air killed germs, prevented eyestrain, and built resistance to disease. Dr. Haven Emerson, who testified before the body, put the importance of natural light plainly: it was no less vital to animals than vegetables.[23]

Reformers articulated a set of concerns that linked good health with good light. They claimed that nobody should have to endure the dark holes that served as rear-tenement homes, to walk through canyon-like streets, or to toil away in interior offices far from bright windows. With the benefits of good light growing ever more evident, it was unsurprising that residents were starting to pay up: natural light had become a commodity. In an interview by the Tenement House Commission, longtime tenement resident Mrs. J. A. Miller quipped that she could find no primary trouble with the slum dwelling. To her, "It seems all about trouble." She said that renters could find a relatively nice corner apartment but it came at a cost: "$2 or $3 more because it is light." In *A Ten Years War*, Riis provided a more complete assessment. Having knocked a hole in the wall of one of his units, a greedy entrepreneur charged an additional fifty cents, "six dollars for this flat, six and a half for the one with the hole in the wall. Six dollars a year per ray." That was the case elsewhere on the block. And everywhere, according to Riis, corner flats rented for more: one four-room apartment with good sunlight commanded seventeen dollars, while a rear one, three rooms but better in every way except that it was dark, was worth only eleven. In troubling tenement environments with limited light and air, it only made sense that landlords commanded higher rents for prized brighter apartments. Such thinking, however, did not sit well with reformers, who contended that light was better thought of as a matter of rights than of dollars and cents.[24]

Fifth Avenue and Wall Street's primary concerns ranged far beyond the standard tenement reform fare. For that quarter, money was more often the issue, and the question was less about the added cost to renters than the fiscal dangers to owners. Worried investors claimed that the outmoded convention whereby property extended vertically from the edge of a lot into the sky could no longer

hold. With light traveling at an angle from sun to earth and wind blowing cross-
wise, one landlord's use of his lot could prevent similar use by his neighbor. Tall
buildings might pay well at first, but before long, when surrounded by similar
developments, darkness came, vacancies rose, and values fell. Financial institutions and realtors agreed: the city suffered with unrestrained development in
part because unhealthy and unnatural conditions undermined buildings' rent-
ability and torpedoed profits. Lawson Purdy, president of the New York Department of Taxes and Assessments, told of skyscrapers halved in value in a handful of years because new neighbors stole light and air. Purdy concluded that
in three to five years, promising developments consistently became commercial failures.[25]

With their critiques of urban canyon lands and congested tenement districts,
Americans brought together a fear of darkness with a love of pure, blowing,
air. Lightness was an abiding concern, and it had its own unique and beneficial
properties, but the same hulking buildings that blocked windows bred stagnant
indoor air. To reformers, sunlight was clearly essential to a healthy and moral
city, but light and air together would become the rallying cry for people hoping
to resolve the city's problems.

It was only a matter of time before legislators attempted drastic remedies.
Early in the twentieth century, New Yorkers initiated bold and well-meaning,
but ultimately limited, endeavors. Though these actions were notable for their
aggressiveness and innovativeness, it is important not to overstate the case. The
city took action within a broader context, and its reformers were well aware
that other places, national and international, had lessons to teach. In *The Tenement House Problem*, Lawrence Veiller wrote an article that summarized extant
legislation aimed at improving conditions for residents. Though somewhat instructive, no American measures were sufficient to stem the tide of darkness. In
contrast, the experience of Europe, especially England, suggested more promising prospects for remedying urban ills. Tenement house exhibits and volumes
on the needs of poor workers took note of the nation in which cities had—
and exercised—the right to bulldoze slums provided replacement residences
could house at least half of those displaced. At times, the European vision was
even bigger, extending as far as municipally funded tenements for the working poor.[26]

Studies found, however, that European models were only somewhat appli-
cable to New York. The old world's slums, marked by squat, overpacked homes,
permitted a tack impossible in a city where six-story buildings teemed with
residents. In dense New York, low-rise rehousing was untenable. Nor would

Americans stomach government expenditures on the scale necessary for municipally funded apartment buildings. That left two options. Model tenements, little more than commercial endeavors by right-minded businessmen willing to accept modest returns, promised tenants healthful conditions. Of course, with sunlight a critical part of good housing, these structures, proponents promised, would offer ample windows and bright rooms. Elgin R. Gould's tremendous special report for the Department of Labor, *Housing Conditions of Working People*, found lots of European and a few American companies willing to do model-tenement work. Gould responded by establishing the City and Suburban Homes Company, broadening New York's effort significantly.[27]

Reformers were enthusiastic about model tenements, but experts like Veiller knew that housing demand was simply too great. They advocated an alternative: restrictive legislation. Americans had tried to constrain what builders did with their lots many times but with limited success. European cities provided examples of much stricter measures, which aggressively capped tenement heights and ensured that yards did not shrink too small. In crafting their 1901 law, New York's reformers took note of the more extreme regulations in place abroad and plotted an aggressive course. They opened up blocks and added space around buildings, requiring that all new structures occupied a more modest percentage of their lot, making hard-and-fast rules for yard sizes, and drastically restricting rear tenements. They also required windows in hallways and, most notably, eliminated the dark room. Every new tenement room had to have windows opening onto streets or courts, and if courts were too small, onto better lit rooms also. Even preexisting tenements did not escape regulation. Landlords had to punch holes in interior walls and tear up facades.[28]

But Progressives worried about the future, not just the present. They feared that the urban problems dramatized so absolutely—and frightfully—in New York threatened the rest of the country. In this case, prevention made far more sense than cure; it was time to take a closer look at some of the nation's other cities and sound the alarm about spreading darkness.

In 1900, Jacob Riis brought his message to Chicago in a series of lengthy articles published in the *Chicago Daily Tribune*. "The Story of the Slum" told readers that ancient Rome, like modern New York, struggled with a tenement problem. Hemmed in by city walls and plagued by gross overpopulation, it built upward. That was a bad choice. Rome fell in no small measure because of the social discord that housing injustices fostered. Chicago was at a crossroads, but its advantage was twofold. It knew the terrible future of a tenement city, and it did not share Rome's walls or Gotham's rivers; unbounded prairie could prevent the old "rookeries," hotbeds of disease marked by poor sanitation, bad air, and darkness. The city had room to grow.[29]

When Jane Addams reported on the findings of a 1901 investigative committee, she acknowledged that Chicago was better off than New York, but she did not share Riis's enthusiasm: it appeared, she argued, "that the slum now building is likely to repeat the history of those in other cities," where it had cost thousands of lives and millions of dollars. There were two culprits in her view: municipal government had abdicated its responsibility to ensure healthy living conditions, and landlords had sought profits with little regard for the "sufficient provision of light and ventilation." Lots were packed and new dumbbell buildings were proliferating, "enveloping the West as yesterday [they] blackened the East." A special investigation found that nearly half of all rooms surveyed in Chicago were dark, very dark, or gloomy. Extrapolating that sample to the entire city yielded 18,000 such rooms housing 22,500 in "a more or less unhealthful condition." The rhetoric of this part of the report, however, was tamer than it had been earlier when the situation sounded much more like New York's. In Chicago, sunlight had become a commodity, which landlords hoarded at the expense of their neighbors, exacting payment in tenants' lives:

> It is the private ownership of the rays of the sun and the health-giving properties of the air. A landlord who builds a tenement to the limits of the lot and several stories high takes from his neighbors both air and sunshine. He also provides many of his own tenants with dark and foul homes. . . . The airless and sunless rooms nourish disease and germs. Babies, almost like blind fish inhabiting sunless caves, suffer from phthalmia. Tuberculosis thrives, and cannot be stamped out without the aid of sunshine. . . . People cannot live without air and sunshine, and strange as it may appear that anyone should have to plead for these things, this Committee and all other tenement-house committees exist pre-eminently for this one purpose.[30]

It was not simply America's largest cities that had a problem. Perhaps it made sense that Chicago was worried that it would grow into New York, but what did the entire state of Wisconsin have to fear? In 1907, it passed America's most sweeping tenement house law, a measure that applied to all cities and towns. Its substance was very similar to New York's, with sanitary requirements and provisions for adequately bright yards. A year later, Wisconsin's Supreme Court determined that the broad and bold article of legislation was unconstitutional because it lacked texture. It wrongly imported many measures from New York and only modified others slightly, often to make them more stringent: Milwaukee was not New York, and semirural Wisconsin was not Milwaukee. To apply one standard universally was a mistake.[31]

This was the background for volume 19 of the *Comparative Legislation Bulletin*, a publication of the State Library Commission in cooperation with the University of Wisconsin's Political Science Department. The report looked broadly at American cities to figure out what legislators were willing to pass and what courts found themselves able to accept. It uncovered a sea of legislation. Connecticut, New Jersey, and Pennsylvania had state laws that applied to their larger cities. Elsewhere, municipalities acted on their own. Though measures took unique forms tailored to localities, the survey found that nearly all legislation required windows opening onto streets, yards, or courts in every room except water closets and bathrooms.[32]

Many other cities wanted to avoid the type of blanket legislation that got Wisconsin into trouble. With eyes toward well-formed and well-informed laws, civic improvement organizations worked alongside government to survey conditions with the ultimate goal of fixing their slums, and they found major causes for concern. Philadelphia, celebrated by many for its single-family workers' cottages and low buildings, housed large numbers of its citizens in structures abutting narrow courts and alleys that permitted little light or fresh air. There were different misconceptions about conditions in Washington, DC. Everyone knew of its alleys, but *Neglected Neighbors: Stories of Life in the Alleys, Tenements and Shanties of the Nation's Capital* surprised readers with accounts of overstuffed apartment buildings, "human ant hills," where darkness made the healthy ill. Jersey City's teeming immigrants trudged up dangerous, dark stairs and into gloomy rooms. The root of St. Louis's housing evil was the one-room apartment. And although Louisville reformers found that conditions were better there than elsewhere, a study chided the foolish who celebrated its 16.6 percent dark-and-gloomy-room rate. The city's health experts were clear about "the value of sunlight, both as a germ destroyer and as a positive health-giving force." Seeing a profound need, in 1910, Veiller announced the establishment of an association to coordinate the meaningful housing movements that had popped up in twenty cities. The assault on urban ills—in existing slums and in housing yet to be built—was national.[33]

For many reformers, the slums, where adults fell ill and parents raised children away from watchful eyes, were bound to be the center of the fight against darkness. For others, however, there was little reason to restrict efforts to the worst of conditions. They also worried about suffering workers and skyscraper owners who forfeited profits when they lost light. Little could contain the anxious many, and city planning held out a bolder solution than tenement reform, one that could bring everyone light.[34]

Thus, in 1893, Chicago passed an absolute cap on heights, basically forfeiting its skyward race with New York. The new law had roots in a debate begun less than a year and a half before. Medical experts claimed that the sun disinfected. Business interests said that property could not extend from earth to the heavens. With proponents pointing to the role of light in healthful construction and opposition neutralized by an oversupply of rentable space, it was decided that buildings could reach a height of 130 feet and no more. In the next thirty years, that number would seesaw up and down, but the method of limitation never changed. A simple cap, reformers argued, could prevent multiple evils, foremost among them, congestion and fires. While Chicago's reformers clearly considered sunlight an important factor in their legislative debate, its role was more prominent for New Yorkers.[35]

In 1890s New York, it was incumbent upon owners to take action to protect their own light and air, and they did. By buying nearby lots, developers eliminated the chance of tall neighbors and thereby protected their own investments. The *Times* conjectured that, eventually, self-regulation would cure the city's ills. Skyscrapers were once perfectly acceptable novelties, reaching high for a little piece of the sky and decongesting overcrowded buildings. But with the proliferation of bulky towers, the old problems of the dark, cramped city had returned in an exacerbated form: "Hundreds if not thousands of once pleasant rooms have become dark and stuffy cells, which, if occupied at all, deserve the immediate attention of the Board of Health, so utterly and dangerously unsuitable are they for prolonged human occupancy." Owners, aware that tall structures cast long shadows and bulky buildings choked out light, worried about future profits and grew to distrust their neighbors. In time, the article concluded, self-interest would dictate self-restraint, and builders would develop only a fraction of their land.[36]

In 1916, with the *Times*'s voluntarism a clear failure, New York gave up on self-restraint and zoned the entire city, but first it took stock of existing legislation. The *Report of the Heights of Buildings Commission* compared restrictions in place nationwide and found that other American cities had followed Chicago's lead and instituted absolute caps. The appendices to the report, however, paid special attention to Europe. Appendix 4 featured five cities, two of them European, and appendix 3 discussed the most effective planning in the world, Germany's zoning system. While each of its cities passed its own restrictions and some stipulations were incredibly complex, in general, Germany treated municipalities as conglomerations of different neighborhoods dominated by buildings of similar girth, height, and function. Smart planning acknowledged that reality with localized prescriptions, which allowed bustling downtowns to grow dense while residential districts remained relatively tranquil. Throughout this work, according to the report, planners prioritized "regulation in favor of light, air and sun."[37]

In the end, New York followed Germany's lead and enacted America's first comprehensive zoning law. The ordinance districted according to use, determining where commerce, industry, and residences would go; area, prescribing the maximum amount of a lot that builders could cover; and height, regulating the shape and size of towers. In that last category, New York City most clearly evinced its love of light and, for that matter, height. Instead of an absolute cap like Chicago's—and the handful of measures it inspired—New York reformers proffered a more elegant solution. Buildings, as had been the fashion in Europe, could grow to a height determined by the width of the street. In a district with a 1.5 multiplier, structures could stand ninety feet tall on sixty-foot-wide streets. But that restriction held only if a building did not step back from the lot line. For each foot back, it could rise three feet higher. Other districts had different multipliers, but for all, a tower that covered only 25 percent of the lot could rise to any height. London's law, too, had permitted setbacks on narrower streets, but it added absolute height caps, thereby checking towers. With its setback regulations, New York looked to allow tall buildings that would not cast big, broad shadows.[38]

For the next forty-five years, setbacks were the norm in New York, meaning that no skyscraper would replicate the blocky silhouette of the controversial Equitable Building. Of course New York did not suddenly become sunny all over, but light most certainly reshaped parts of the city. Relatively underdeveloped Queens would not change much, and most of Brooklyn was more suburban than urban, but apartment buildings the city over would cover less of their lots, and New York's skyscraper architecture assumed its distinctive shape, carved in no small part by light. The Empire State Building's tremendous tower impresses, but its wide base gives way to a series of setbacks. The Chrysler Building is similarly designed. Indeed, sunlight sculpted New York's most distinctive architecture and refashioned some of its most noted districts.

As had been the case with tenement reform, the zoning movement spread, and though Chicago gave the first small nudge toward comprehensive urban plans, New York really made the case. In 1917, Edward M. Bassett, whose legal

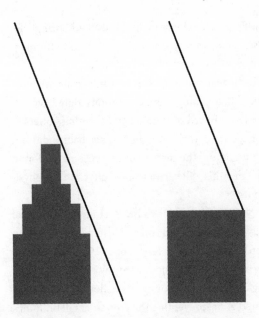

Figure 1.2. Light blocked by buildings designed with (left) and without (right) setbacks. Diagram by author.

arguments helped give form and substance to the planning debate, claimed that the future looked bright for his movement. Courts, he reported, had permitted cities to restrict the ways people used their land, provided legislation was reasonable and necessary to preserve the public's health and welfare. Zoning laws fit both these requirements.[39]

Four years later, a representative of Pittsburgh told the Thirteenth National Conference on City Planning that Pennsylvania had passed an act authorizing cities of the second class to enact zoning ordinances, which would restrict height, use, and lot coverage. And five years after Pennsylvania took action, a revised pamphlet published by the Department of Commerce noted that by the end of 1923, 218 municipalities housing twenty-two million people had zoning restrictions in place. The publication, however, was no mere statistical accounting; it was a standardized enabling act by which states could empower their cities to district in a legally defensible manner. Though the substance of the model was intentionally vague, permitting all manner of restriction, its reasoning was not: urban districting was defensible for its effect on health and morals.[40]

The legislation that emerged across the United States assumed widely varying forms, and some cities showed decidedly little interest in sunlight, but for others—including some hardly overbuilt locales—this was not the case. In city planning commission reports and studies favoring legislation published by

El Paso, Memphis, Akron, San Francisco, Houston, St. Louis, and Springfield, Massachusetts, the light needs of cities were a part of the justification for planning. Other places made their arguments particularly forcefully.[41]

Milwaukee, a bastion of reform zeal and one of American socialism's most significant success stories, argued that zoning was a community right. According to a tentative report of the city's Board of Public Land Commissioners in 1920, it was "common knowledge" that without light and air, babies did not grow properly. Smart and bright zoning, the report concluded, "makes for an orderly city and it can be shown that this will have a marked effect on the physical fitness of the city's inhabitants."[42]

Farther east, Newark took New York's lessons to heart, chiding its colossal neighbor for ignoring the needs of its citizens and abdicating its responsibility to provide ample sunshine. According to a Newark report of 1919, the 1916 zoning law was inadequate, leaving builders little choice but to pay each other not to construct light-and-air-stealing buildings. The younger, smaller city would not make the same mistakes; it would zone, preventing the growth of tall buildings that blocked neighbors' light and attending to the minutiae of sun needs: the differences between light in winter and in summer and between north-south and east-west exposures.[43]

Legal challenges to planning legislation were common; they attacked all parts of the laws. Courts, while refusing to say that cities and states had absolute rights to plan, tended to uphold the new measures. In New York, the legislation and its ability to withstand legal challenges were predictable—the city had an obvious problem. But other cities had far less reason to complain. Nationwide, laws and litigation came; Maryland's appellate court had to weigh in (1927), as did the supreme courts of Washington State (1920), New Hampshire (1928), and Wyoming (1934). In each case, the judiciary upheld the zoning measure as a just use of the police powers, in part, because of the relationship of light to health and the public's welfare.[44]

City planners and tenement reformers agreed that darkness was a problem. It was a cause of disease, a moral contagion, and a source of economic instability. It killed, corrupted, and impoverished. Municipalities, authorities argued, needed better and more programmatic planning. But theirs were not the only voices. In the first quarter of the century, reformers in schools and hospitals spoke out too. Though there was cause for worry about people of all ages, many efforts paid special attention to the needs of America's fragile young.

Publication 116 of the New York Association for Improving the Conditions of the Poor discussed the Home Hospital, a new form of treatment facility for tuberculosis. Rather than breaking families apart and taking the sick away, the hospital kept a hundred residents, parents and children, well and ill, together in a healthy environment intended to heal the patients and prevent everyone else from succumbing to infection. The Home Hospital's design gave to its resident patients and their families all the light they needed to get better: large windows, small balconies, and on the roof, a park. The publication—really more of a brochure—showed patients enjoying naps outside in the sunshine and inside brightly lit rooms. Within four years of opening, the hospital was a great success and had twice doubled in size.[45]

The Home Hospital was not the only New York medical facility interested in bringing light to its patients. In the first decade of the twentieth century, the *Times* often described hospitals' prospects for brightness: a new sunhouse for the Presbyterian; sun parlors for Staten Island's new tuberculosis hospital; and all-light wards, plentiful balconies, and rooftop gardens for Bellevue. Elsewhere, the therapy met with even greater enthusiasm. The role of nature in tuberculosis treatment was not new. In the nineteenth and early twentieth centuries, western and southern boosters claimed for their regions an unparalleled role in the treatment of suffering easterners. Fresh, bracing air and a healthful lifestyle awaited those willing to take advantage of nature's bounty. Some did, heading for Los Angeles, Denver, or Tucson, although the facilities they checked into were more like small rest homes than the large, bureaucratically organized institutions that grew prominent in the twentieth century.[46]

Other hospitals represented a more individualized approach and on a drastically different scale. Dr. John Harvey Kellogg's Battle Creek sanitarium was one such place. Despite its notoriety, the facility was far from unique—not even the oldest of its type. In the 1870s, Edward Trudeau had founded a similar institution; however, Kellogg was exceptional for his impact on care facilities and on popular thinking about health. The surgeon, who received his conventional medical training at Bellevue, had moved far from orthodoxy. He believed that corporeal impurities plagued the lives of modern Americans. But the ills of civilization could be remedied: assorted dietary choices—abstinence from alcohol, vegetarianism, lots of fiber (his brother Will went on to found the Kellogg Cereal Company), and thorough mastication—would help the body flush toxins and, in doing so, promote health. Kellogg also regarded light as a natural tonic and an essential antibiotic. In his book *Light Therapy: A Practical Manual of Phototherapy for the Student and the Practitioner*, he touted this new "physical therapeutic," which cleansed the body, cured metabolic disturbances, eliminated anemia, and laid tuberculosis low. With memberships in the British

Gynaecological Society and France's Société d'Hygiène, Kellogg claimed that he was part of a strong international community of practitioners. Light baths, he said, were born in Battle Creek, grew prominent in Germany, and returned to an enthusiastic American audience. Contradicting Kellogg's claims of a strong international and American contingent of light therapists, his work only achieved widespread acceptance in the States when doctors elaborated upon it in later years. Those future experts, though, would not sanction some of his findings, foremost among them, his belief that heat, as much as light, healed and that incandescent lightbulbs were good for health.[47]

With its growing emphasis on natural light, hospital design fit nicely with prevailing beliefs about disease, but it was limited in that facilities only treated the infirm. The school offered a different kind of space, one that could bring untold benefits to young people nationwide. It could teach them good habits, help them understand the merits of sunshine, and show the perils of childhood wasted.

Constance D'Arcy Mackay's *Pageant of Sunshine and Shadow* (also called *The Child Labor Pageant*) used the school play to bring the scourge of child labor into the light. It features newsboys, children in sweatshops, and small laborers behind pushcarts, at shoeshine stands, under miners' caps—scene after scene of "toil" and "shattered childhood." The misery pauses in the sixth scene, which takes place far from a city or industry in a wooded "Court of Happiness," where girls perform a "Russian dance, in Russian peasant costume . . . and a Swedish dance in Swedish costume." The subsequent episodes volley back and forth between "the Joys of Childhood" and "scene[s] of continued labor." In the end, MacKay leaves her audience on a happy note, with "the spirit of Youth" appearing "her arms outstretched in pleading for the betrayed youth of the world." The trope, children are safe in light and suffer hidden in shadows, was one with which MacKay's audience was no doubt familiar.[48]

On June 5, 1916, reformer Lewis Hine went to New York City's Washington Irving High School to shoot photos of performers in a production of the MacKay pageant on their building's rooftop. Just over twenty girls posed smiling. Though Hine's subjects put their school's roof to special use, they were not alone atop the city. In 1895, the city passed a law that each school had to have an open-air playground proportional to the size of the building. For some schools, that could mean space at ground level. For many, there was no place to look but up. Between 1898 and 1902 and then again in 1907, the annual report of the Department of Education compiled a list of its buildings under construction and additions to preexisting structures. With an increased municipal commitment to education, this was a time of great growth for the school system. Over the period, there were fifty-eight projects, and a quarter of them, ministering to thousands of kids, featured rooftop playgrounds.[49]

In McKay's work, sunlight was primarily a metaphor, and the rooftop playground, while a boon for the light it brought, was as much about exercise and fresh air as anything; however, C. B. J Snyder's plan for the structure of schools more clearly indicated a commitment to sunlight. Snyder, the city's superintendent of school buildings, wanted to reshape the schoolhouse entirely. His vision might be described as the anti-dumbbell; rather than a fat middle section, the H-shaped school would have a slender middle that connected two long wings on the sides of the lot. The intent was to create large courtyards in front and back and, according to Snyder, "secure an abundance of light and air to each class room." True to his word and confident in his innovation, Snyder pushed hard for his design, with roughly half of the new constructions built around the turn of the century H-shaped. The forward-thinking work of this reformer brought him Riis's praise: he was a man "who found barracks where he is leaving palaces to the people." Though Riis was guilty of hyperbole, his admiration was genuine for buildings he described as masterstrokes, inspired by Parisian architecture and copies of "the handsomest of French palaces, the Hôtel de Cluny," with "not a dark corner in the whole structure, from the splendid gymnasium under the red tiled roof to the indoor playground on the street floor, which, when thrown into one with the two yards that lie enclosed in the arms of the H, give the children nearly an acre of asphalted floor to romp on from street to street."[50]

New York's altars to the letter H and rooftop playgrounds were not the full extent of sunlit education. With the fresh air school, popularized in the nineteen teens, reformers sought a broader program that would touch countless youngsters' lives. Afraid that an unsocialized and un-Americanized rabble was overtaking the city, school reformers of many types sought an educational program that would teach civility and discipline. Recent scholars have revised the conventional historical portrait of these heavy-handed activists, pointing out that parents and students too shaped the course of instruction and play, but little of this work treats schools as physical spaces that children enjoyed—or endured.[51] In a sense, the problem of the city and the problem of the school were the same. Rooms that educated children, like rooms that housed them, could sicken. The struggle for the schoolhouse, however, was different from the battle against the slum; reformers worried that when parents closed their doors, they did whatever they wanted. In the school, instructors were in control. There, the physical space could unite with a healthful daily program.[52]

The fresh air school was less ambiguous in its goals than the rooftop playground. The prevailing benefit to children was not play, nor did it have much to do with classroom learning—though kids did learn. Rather, improvement in the physical environment was seen as a good in and of itself. New York reformers had long before established programs with the goal of getting children out of the cramped city and into an environment more suitable to their growth. In cities big and small, this new school movement sought far more, to bring nature to students with classrooms better planned for health.

Louise Goldsberry occupied the often-narrow border between enthusiast and fanatic. She set out around 1910 to document the world's open-air schools and outdoor education programs with a study that made up in scope what it lacked in rigor. Her product, however, was more than a simple reflection of one woman's hard work. Goldsberry sent out letters requesting information, especially photographs, to superintendents of education, state tuberculosis and public health associations, and private schools around the world. Her correspondence, completed over more than five years, bared considerable fruit: twenty-five hundred pictures and a wealth of information. As a supplement to her study, Goldsberry provided a manuscript to help explain what she found and to fill out the history and bright future for her beloved school movement. Her work, therefore, represented not just one woman's take on reform, but evidence of what these schools and institutions hoped to show of themselves.

The institutions Goldsberry celebrated sought to ameliorate the conditions that crippled children, from environmental stressors like darkness and bad air to ignorance of the "very laws of life and health." Her photos documented 343 programs in 184 locations, mostly urban, in forty-one states. The collection, its comprehensiveness, and Goldsberry's attempt to represent the full geographical scope of fresh air work reinforced her message. A young and, she hoped to show, growing movement, fresh air schoolwork held great promise. The programs in her photos were diverse but each was systematically planned and deliberately overseen. Though many existing institutions served only children, who reformers believed most needed help, her goal was clearly more expansive, and her mound of photographs made as much clear. Fresh air schools belonged everywhere.[53]

Goldsberry trumpeted her ability to get responses from school administrators in the Central Powers before the First World War began. Indeed, that was a minor coup, because Prussia, Hungary, and Bavaria had some of the world's very first forest schools. In general, work on the Continent demonstrated a deeper embrace of sunlight than was yet common in America. At Auguste Rollier's alpine clinic and outdoor school in Leysin, Switzerland, destined to be-

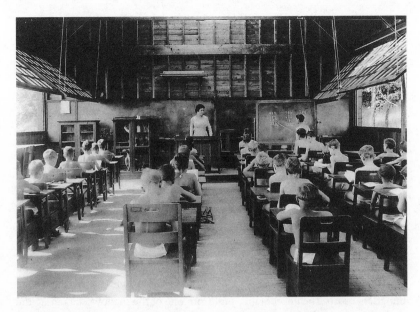

Figure 1.3. Schoolhouse, Perrysburg, New York, ca. 1900–1920. Box 42, Goldsberry Collection of Open-Air School Photographs. Courtesy of the Library of Congress.

come a model for American institutions, children were "a perfect portrayal that needs no phrases to talk its peerless work," "each a book for reading—of open-air and out-door cure and happiness." In Goldsberry's text, the children were naked, but in the actual photographs, Rollier's deeply tanned patients were always adorned in at least a loin cloth, if never much more, whether sitting, standing, studying, playing, or even taking their lessons in a sunny alpine snow field. Attire was similar in another Swiss clinic, while children in German and Austrian facilities wore only slightly more.[54]

In America, there was only one facility where children dressed as they did in Switzerland. By the nineteen teens, Battle Creek was no longer the nation's hot spot for light therapy, having lost its supremacy to the J. N. Adams sanitarium and school in Perrysburg, New York. There well-tanned children passed their days in little clothing while receiving their education. According to Goldsberry, this was America's best "analogy" to Rollier's work, a place where "children are given over, naked, to nature's sun and snow and summers," and where "roaming in the winter's white bleakness" and "in the summer's glow" were "sun-healing."[55] These institutions were intriguing exceptions in the fresh air movement. Very few American schools sought deeply bronzed children or ruddy young outdoorsmen. Generally, in the decades after the turn of the century, healthy skin was still pale skin. Most of Goldsberry's photographs showed schools akin to those in England and Scotland, more conventional facilities, where children got a bit of exercise outside, took a lesson in the open, or studied in a schoolhouse modified for health.

With children in need of a complete program that kept them well, Goldsberry's project turned up institutions almost as likely to emphasize rest and recess as learning. At Berkeley's summer play schools, the young children sat outside in dappled sunshine while older ones exercised in the open. In one image, trees ring a field, casting shadows—shade the children have taken effort to avoid, with all but a few of them well spaced in the clearing and perfectly placed in the light. New York's photos most effectively testified to the sun's role in the fresh air school. A 1918 picture of the Federation for Child Study shows a group of children dressed in white holding hands in a space surrounded by buildings. Healthy and active, they seem to radiate the bright sunshine their young bodies hungrily absorb. The photograph, shot from one of the shadows darkening the border of the playground, depicts an ideal of healthy, urban recreation.[56]

Besides play, it was study and nap time that captured the most attention in Goldsberry's photo collection; since both of those activities occurred in classrooms, the pictures are often similar. They tend to show sunshine streaming into the students' work and rest spaces. With conditions lighter outdoors, the areas closest to the windows are brightest. The contrast, however, rather than making the far side of the room seem grim, accentuates the cheerful and healthful conditions. In some pictures, an entire wall of windows, washed out from the overexposure necessary to properly light the rest of the photo, creates a gloriously blinding effect with no building visible on the other side of the overlit glass.[57]

Fresh air rooms were often celebrations of windows, with glass ranging from just above the floor to tall ceilings, open wide or glowing from the sun's light. Many structures were freestanding and covered with windows, permitting four sides of exposure. A Pennsylvania tuberculosis preventorium had walls almost entirely of glass,[58] and a New Orleans institution had a similar physical plant. But neither of those buildings, nor Milwaukee's much larger facility, could compete with Moraine Park Day School in Dayton, Ohio. With its pitched roof and nearly all-glass construction, "the school in the glass house" looked more like a greenhouse than a school and was, most of all, a celebration of sunlight.[59]

One institution in New York, the ferryboat school, had no need for windows at all. Ferryboat schools probably helped with the overcrowding of classrooms, but they also gave children lessons in bright sunshine. One photograph offers

Figure 1.4. Children playing, 1918. From the Federation for Child Study, New York City. Box 45, folder 2, Goldsberry Collection of Open-Air School Photographs. Courtesy of the Library of Congress.

a beautiful view of the city in the distant background, the Brooklyn Bridge and downtown Manhattan barely visible in a sunlit haze. The lesson takes place under cover from an upper deck; still, the sun streams in, bisecting a chalkboard and brightening most of the scene. More commonly, rooftops provided impromptu fresh air spaces. A Richmond, Virginia, photograph captures a primitive program—little more than a couple of nappers on cots on a roof. In an image at New York's Public School 21, a large group of youngsters recline on a deck, three stories above street level. Above them rise two walls of the school building, but the children enjoy their rest out of the shadows. Elsewhere, tent schoolrooms, sometimes atop city roofs and sometimes on the ground, permitted teachers and administrators the flexibility to retract cloth paneled roofs or tie back fabric walls in good weather.[60]

The staged photographs of students behind empty desks on wooden floors in tented, spartan rooms often reveal reformers' love of order in addition to their quest for light, but one from an Ohio school more assertively conveys different messages: purity and civility. The well-dressed students sit up straight. A cloth ceiling is rendered translucent, and the scene behind the unenclosed back

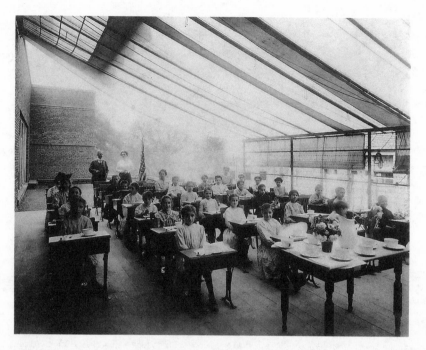

Figure 1.5. Murray Hill School, Ohio, ca. 1900–1920. Box 54, Goldsberry Collection of Open-Air School Photographs. Courtesy of the Library of Congress.

of the "room" is a bright haze. Ceramic teacups, deliberately placed on children's desks, transmit a sense of propriety; the flowers on the teacher's, beauty. The scene is a perfect counterpoint to Riis's squalid tenement halls and frightening alleys: orderly, clean, cheery, and above all bright. Light, as Progressives long assumed it would, has cleansed the scene.[61]

In reality, conditions for students were probably less wondrous than the sunny photographs suggest. Fresh air schools might have been comfortable in the spring, but in wintertime, they must have been brutal. Photographs from Hartford, Chicago, and Kenosha, Wisconsin, show young children sitting at desks or resting on cots, bundled up with hats on and wrapped in blankets. The expressions on their little faces vary widely, from cheerful smiles to dour grimaces and stony stares. One of Chicago's sons, Tony, shows off his "Eskimo suit," the city's standard cold weather outfit: a thick wool hooded coat and pants, with knee-high insulated boots. He stands in the snow, smiling, though his expression looks more than a little forced.[62]

While it would be a mistake to assume all institutions were alike, it would

Figure 1.6. Chicago tent school. Photograph by F. P. Burke, ca. 1900–1920. No. 224, Goldsberry Collection of Open-Air School Photographs. Courtesy of the Library of Congress.

Figure 1.7. "An open air class on *Day Camp Rutherford*, New York (Manhattan Island) across river," 1911. Photo from Bureau of Charities, Brooklyn, New York. Goldsberry Collection of Open-Air School Photographs. Courtesy of the Library of Congress.

Figure 1.8. "Anaemic Class," Public School 21, New York City, ca. 1900–1920. Box 45, folder 2, Goldsberry Collection of Open-Air School Photographs. Courtesy of the Library of Congress.

be even less accurate to consider these programs haphazard. Treatments were often quite similar. The Eskimo suits in Chicago were like those in St. Louis and similar to the sacklike outerwear in New York. Richmond rooftop loungers could have been in Chelsea, Massachusetts, or Reading, Pennsylvania. Tent schools in Brooklyn mirrored those in Chicago. One image from an Oregon school looks a lot like a classroom in New York City, where study time was probably different, but the rest period was not. The rooms, twenty-five-hundred miles apart, had

Figure 1.9. Fresh air class, rest hour, Public School 51, New York City, 1911. Goldsberry Collection of Open-Air School Photographs. Courtesy of the Library of Congress.

desks pushed out of the way, children in sleeping bags, and nearly identical windows, which pivoted in their middles for maximum exposure to the elements.[63]

These buildings differed significantly from workplaces, which also experimented with new styles of windows and construction. Generally, employers—factory owners or office managers—sought ways to light their workplaces more effectively without any added cost; for them, sunshine was an economic choice, a way of getting for free what would otherwise cost money. In the case of the fresh air schools, the objective was different, a healthier environment that could help students overcome the malevolent city. If ventilation was the first premise of the new educational regimen, it was not the only one. Air could dilute the density of microbes in a room, but with a sense that sunlight mattered to the welfare of their young, reformers seized upon new patterns of construction and an innovative school program to help their charges grow.

By the nineteen teens, sunlight was a clear concern for urban Americans. They feared that New York's present would become the nation's future. Businessmen

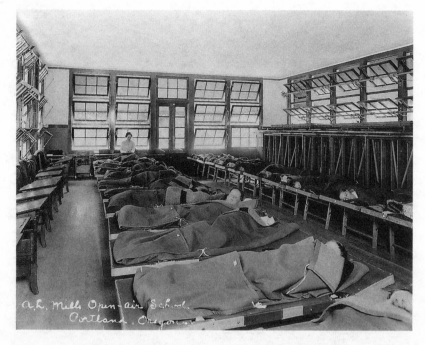

Figure 1.10. A. L. Mills School, Multnomah County, Oregon, ca. 1900–1920. Box 58, Goldsberry Collection of Open-Air School Photographs. Courtesy of the Library of Congress.

had decided that sunshine was worth paying for, and natural light now came at a premium. Reformers, unhappy with this new reality, looked to secure a more egalitarian and brighter city, which ensured investments' profitability while securing at least a minimum of health for all. In schools, they determined, children had basic needs that were not met in the average classroom. Better-designed buildings sporting glass walls and flooded with the sun's rays could help.

Social reforms dominated this early sunlight thinking: unmake the slum, redesign the school, reshape the city. But the popularity of those fixes would not last long. By the 1930s, securing sunlight had become the responsibility of the individual, and scientists, not planners, were providing the necessary tools. Gone too was the hazy thinking that predominated earlier: Kellogg's incandescent bulbs were worth little, Perrysburg's tanned youngsters were far from exceptional, Guerard's constitutional medicine was mainstream, businessmen used sunlight to promote worker health, and sunshine was a force for wellness quite distinct from all other environmental factors.

2

The Dawn of Scientific Sunlight

In 1929, the National Carbon Company, a division of Union Carbide, published its "Sunshine Map." Showing the entire country, it gave graphs for about fifty cities, which compared how many hours of sunlight were possible each month to how many actually shined after clouds had their obscuring effect. The goal was to dramatize light lost throughout the year. That objective made sense for a company that produced the Eveready Sunshine Lamp. But Union Carbide was not alone in associating gray skies with limited light, and clouds were not the full extent of the complications for cities hoping to overcome their darkness problems.[1]

The 1920s began with great hope for brighter cities. Planners, doctors, and school designers would organize metropolises and, in so doing, cure urban ills. By 1925, such solutions to sunlight problems had lost their luster. A new set of concerns emerged, born of new science. Two main discoveries—rickets and the importance of ultraviolet light—changed discussions about sunlight's uses and torpedoed many hopes for healthy homes through smart urban planning. All of a sudden, it appeared, glass could not save people from the "diseases of darkness," foremost among them rickets and the dreaded scourge tuberculosis, for glass, like pollution, filtered ultraviolet rays, critical for good health. Rooftop playgrounds, setback architecture, and big courtyards could do little for a city in which children played under smog-filled skies and people toiled away indoors. With time, the news grew even worse. Research would show that even if the sun did get through to suffering urbanites, the prognosis for them could still remain bleak.

Fortunately, innovation and industry soon came together to provide people with what, only a short while earlier, they had not known they needed. Inventions came quickly: new types of glass with different transmission properties and lamps that simulated sunshine. Suddenly, there were hosts of commercial products promising nature's great gift. The new knowledge of sunlight, with its accompanying critique of urban life, embraced not an idealized past but a brighter future in which technology brought salvation. With a stamp of scientific sanction, pitchmen brought word of these healing inventions to the public. In doing so, they helped to replace a broad public health movement, which sought reform for entire cities, with a more individual approach. That some could not afford luxuries like sunlamps may have been regrettable, but the businesses that made a profit on new products seemed little concerned.[2]

There was nothing new about rickets. Doctors had, for some time, been aware of the affliction, which gruesomely deformed children and had consequences that lasted well into adulthood. Victims suffered from soft bones, so soft, in fact, that they bent under the weight of the body, yielding the telltale signs of a rachitic childhood: bowed legs and arched backs. In very extreme cases, doctors found, arms might stretch, pulled down by gravity. The deformities stopped getting worse after early childhood, but by then, the damage had been done; rachitic children were crippled for life. According to best medical practice, less extreme cases left youngsters relatively normal in appearance but weaker and more susceptible to infection. While the signs of severe rickets were well known in 1915, its etiology was not.

The road to understanding was winding. While working as a missionary and doctor in Japan, Theobald A. Palm discovered that rickets was far less common in Asia than in his native England. In an impressive piece of speculative science, Palm sent letters to missionaries throughout Asia and North Africa, asking them to note rickets rates. He concluded from this semiformal survey that rickets was largely the product of a temperate climate; furthermore, he hypothesized, it was an urban disease, a problem where population was dense and industry robust.

In his 1890 paper, "The Geographic Distribution of Rickets," Palm determined that sunlight was key. Through an unknown—and, as yet, unimaginable—process, the sunlit baby avoided the disease. Palm saw a problem, a rising tide of rickets in his urbanized, industrialized Britain, and he would not stand by inactive—his solution, a system of preventive medicine, facilities where children could receive therapeutic sunbaths.[3]

Palm's work gained some traction in the years to come, with doctors picking up his thinking in a few medical monographs and some journals, but it would take more science to bring his work fully to the fore. Around 1920, British researcher Edward Mellanby breathed new life into Palm's line of thinking when he found that a dietary deficiency gave rise to the condition and that the likely missing factor was fat-soluble vitamin A. With only three known vitamins (and the other two ruled out for chemical reasons), it seemed a reasonable conclusion. In his paper reporting experiments on two hundred puppies, Mellanby was satisfied that he had guessed right: A, found in animal fats (especially cod liver oil) and in lesser concentration in vegetable oils, cured rickets. Though this seemed the most probable theory, the scientist had to concede that there could have been some mysterious, nearly identical associate of the known vitamin that mattered.[4]

Mellanby's best guess proved wrong. There was indeed another health factor at play, and within a couple of years, his findings were in doubt. E. V. McCollum's team at Johns Hopkins included some of the earliest and most forceful critics. They tested the ability of different types of fats—most of which were high in vitamin A—to prevent rickets. The Englishman had been correct in his assessment that cod liver oil worked wonders for health, but remarkably, successful treatments continued even after experimenters reduced the fishy fat's vitamin A content.[5]

In the early 1920s at Columbia University, Alfred Hess and his laboratory provided the last critical piece of evidence when they validated Palm's work by using sunlight to prevent rickets in infants and rats. Hess's experiments spanned years and expanded on the findings of German scientists, who in the late nineteen teens, had cured the condition with ultraviolet light. By 1924, researchers in three countries on two continents had built a rough version of the case against rickets: some mystery substance, probably distinct from vitamin A—but maybe not—enabled the body to efficiently take in calcium and phosphorous and to deploy it to the bones, facilitating the construction of a strong skeleton.[6] With more evidence, the relative scarcity of this new vitamin, later called D, became clear. Cod liver oil was one of its few great natural sources. Another was the skin, a vitamin factory, needing only energy, which it got in abundant quantities from sunlight, to catalyze its production.[7]

There were countless experiments and many wrong turns before these surprising conclusions. The line of reasoning from scurvy to vitamin C to citrus was relatively simple, the causation clear. With D, that was not the case. It was hard to sort out precisely what went wrong with Mellanby's research, but part of the problem derived from the inherent challenges of understanding a disease without a single, specific cause (vitamin D needed calcium to work its wonders) with two drastically different cures (sunlight and a rare dietary element). Mellanby fed his dogs a handful of basic diets and modified them however he saw fit, creating a host of test conditions with few subjects in each. According to Mc-Collum, that undermined the Englishman's analysis and caused him to underestimate phosphorous and calcium's value. In fact, Mellanby did find that calcium seemed to be of some importance in preventing rickets, but that discovery did not affect his main conclusions substantially. For Hess, some rickets research had suffered from inaccurate results because it lacked careful attention to sunlight; he would only trust experiments that controlled for it.[8]

Hess was only half right; sunshine was indeed potent, but further science would soon prove that all sunlight was not equal and that rats raised in glass laboratories were still susceptible to rickets. "Real" sunlight, it turned out, was far scarcer than tenement reformers thought. The science that uncovered this remarkable fact had roots in findings almost 250 years old. Back in the seventeenth century, Sir Isaac Newton ushered in a new era in optics when he discovered that sunlight was not a single beam but a composite rainbow of colors. Subsequent physicists found that it was composed of waves, which varied according to their length—from longest to shortest: red, orange, yellow, green, blue, indigo, violet. Clouds, made of relatively densely packed molecules of water, scatter all visible wavelengths similarly and appear white. Smaller molecules, however, affect light by absorbing and reemitting waves. The molecules of the Earth's atmosphere are good at taking in and giving off blue, scattering it broadly so that it colors the entire sky. At the end of the day, with the sun close to the horizon, light passes through extra atmosphere, resulting in more scattering of red and orange—the effect, sunset.

This information was no mere curiosity or object of children's wonderings. Around 1800, scientists uncovered some mysterious invisible rays that could catalyze chemical changes. Later, they divided the colorless parts of the solar spectrum into two main categories (themselves subdivided): ultraviolet (shorter than violet) and infrared (longer than red). By the 1920s, researchers

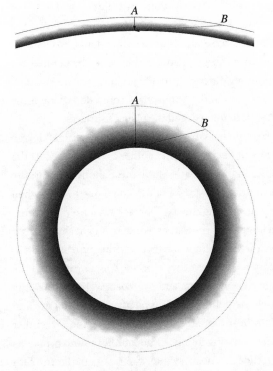

Figure 2.1. The distance traveled of light through atmosphere with the sun straight overhead (A) and just above the horizon (B). Neither image is to scale, but the top one is far closer to accurate (Earth is nearly four thousand miles in radius, and the dense parts of the atmosphere are less than thirty). Notice that as the circles approach each other in diameter, the difference between A and B grows. Diagram courtesy of Drew Youngren.

had become confident that the former portion of the spectrum was responsible for preventing rickets and eliminating bacteria. Downes and Blunt's pioneering experiments that pinpointed blue as the disease killer broke critical ground but were wrong in the wavelengths they privileged. The findings that led to this new understanding were not a cause for celebration. As the story of the setting sun told, not all rays reached the earth with the same ease. Ultraviolet was a broad spectrum of light, with the shorter region, blocked by atmosphere, never striking the planet at all and the longest, most penetrating waves, far less biologically active. The region of sunlight that doctors now believed could lay microbes low and build strong bones was really a narrow band, one that struggled to reach earth under the best of conditions.[9]

Few places, however, had the best of conditions, and geographic and environmental variability complicated sunlight concerns. Skyshine, or cloudshine, as it was occasionally called, was simply sunlight reflected off clouds, but since clouds could actually transform the content of the light, all skyshine was not the same. In 1922, General Electric lighting pioneer Matthew Luckiesh noted that light from the sky, from the sun, and from reflections off buildings had different properties. In subsequent years, others would corroborate and complicate the finding. W. W. Coblentz, of the United States Bureau of Standards, brought the sanction of research science and authority of government when he pointed out that skyshine was fickle, responding differently to different molecules. And doctor Henry Laurens agreed: particles in the atmosphere could determine if the biologically active part of the spectrum reached the planet's surface.[10]

In little time, the full consequences of these discoveries became clearer. Again, Palm was proved right, but he had missed another critical matter. At the equator, the altitude of the sun in the sky varied little throughout the year: the date did not matter; with clear skies, the sun was always high overhead—when light was richest—at some point during the day. Further away from the equator, that was not the case. In temperate-zone cities, winter sunlight shined from close to the horizon and was, therefore, impoverished. Further north, the situation was even direr. It was easy to see that heavy clothing was a problem in the absorption of winter sunlight, but wool hats and overcoats were not the only barriers to health. In an article written for the *Journal of the American Medical Association* (*JAMA*), Charles Sheard found that ultraviolet light in December was only a twentieth as strong as it was in July. Other studies found problems similar in kind if not in magnitude; in Albany, winter sun was a quarter to a fifth as strong as it was in summer, and in Toronto it was only an eighth as potent.[11]

Further scientific studies indicated that the dangers and benefits of a given latitude were not the same for all people. According to many experts, the alpine climate was uniquely suited to the sun-starved. It did not matter as much if light shined from close to the horizon if the air was clear and thin. This thinking made the Swiss Alps the world's great heliotherapy hot spot. It was unfortunate, according to light enthusiasts, that America had not fully explored its own alpine sunlight. With little knowledge of the opportunities within its own borders and scant indigenous hard science about the merits of altitude for sunlight therapy, it had to import data. In one article, Frederick Tisdall reported that England had recently conducted a study that found that the English countryside received about four times as much ultraviolet light as London but only half as much as the Alps.[12]

Doctors concluded that the relative scarcity of sunlight in some locales was a major problem. The indictment of cities grew with the discovery that, while pollution choked out some visible light, it did more damage to the shortest waves. When he traveled to America, British eugenist, smoke-abatement expert, and sunlight authority Caleb Williams Saleeby celebrated new world efforts. Throughout Canada and the United States, he claimed, public health officials had spearheaded pollution ordinances that effectively chased smog away; dark skies were receding, and the typical American city had little to fear. What was more, Saleeby contended, the atypical ones were in far better shape than London: even New York and Boston were little polluted.[13]

If foreign visitors were excited by what they saw, the American evaluation was far dourer. In fact, according to New York's health commissioner, the city lost as much as 45 percent of its noontime sun to smoke. A Chicago report found its pollution a "matter for serious concern" even in summertime, and winter was far worse. Moreover, these disturbing conditions were not limited to the nation's biggest cities or industrial hubs; one report in the *American Journal of Public Health* (*AJPH*) generalized in claiming that there was twice as much ultraviolet radiation in the country as in the city. Newspapers informed the public about the looming threat in reports that assumed technically savvy readers; ultraviolet rays, already understood, generally required little explanation. In one *Washington Post* article, Dorothy Pletcher put the smoke problem clearly, asking, "Isn't it rather appalling to have the American Medical Association tell us that the smoke pollution of our cities is depriving our child life of half to two-thirds or more of these beneficial and necessary rays?" She lamented that a medical hope was being lost: "Science, it would seem, has discovered a great boon to mankind only to see it snatched away immediately."[14]

A well-informed public took note and reached out, seeking a deeper understanding of its environment. In 1924, Eugene H. Smith wrote to Utah's branch of the United States Weather Bureau requesting information about the relative ultraviolet content of sunlight at different altitudes. When he did not get the information he sought, Smith recast his question with greater specificity. He wanted to know how much more ultraviolet light of the wavelength 304 mμ there was at 4,310 feet than at sea level.[15] This was not an eccentric request. From Georgia to Maine, inquisitive citizens sought similar information. The bureau responded as best it could. Publicly, by the late 1920s, it was claiming that the nation had a serious pollution problem, one that ranged beyond dust damaging lungs and dirtying buildings. That was bad, but the problem of blocked ultraviolet rays was worse. Smoke concerns found their way into newspapers, as authors looked for issues of public interest and found this one. In reflecting and fueling concerns about smoky, ultraviolet-deprived cities, the Weather Bureau fostered interest in a study of pollution's effects.[16]

It was little wonder, then, that in 1928, the *New York Times* reported enthusiastically on an upcoming eleven-city smoke study—with locations chosen for geographic diversity, including New York, Salt Lake City, Atlanta, and Lincoln, Nebraska. The article announcing this important work focused readers' attention on the importance of ultraviolet light to America's welfare. Unfortunately, excitement for sound science regarding the ultraviolet crisis mixed up matters and raised expectations the bureau could not meet. Short on funds and the necessary technology, the government could see no way to conduct the study people wanted. The subject it had the means to examine was dust, not sunlight or ultraviolet light.[17]

In lieu of an examination of lost sunshine, the bureau did the best it could. With a little mathematics and data offered by European experts, it provided approximations of the effects of latitude on sunlight content. When Charles J. McCabe of the Detroit Bureau of Smoke Inspection and Abatement wrote asking for information about ultraviolet rays, he found out that, based on European observations, there was little hope for robust light in Detroit; Ralph M. Shaw's factory in Burlington, New Jersey, would not fair much better. Conditions were somewhat more promising for Mrs. Lester G. Seacat of Kansas City, Missouri, but far from ideal.[18]

The connection between clear skies, bright light, and good health allowed for odd claims by unconventional sunshine advocates. The First International Conference on Light convened in 1928 in Auguste Rollier's Leysin sanitarium. There, Saleeby told an impressive international group of authorities:

> Only custom-engendered blindness can regard the demands of primary sanitation as met by the provision of pure water, whilst the air is abominably befouled, and the light of life is blackened by the darkness. . . . For lack of the primal necessity, sunlight, we die in vast numbers far exceeding those due to any other cause.

Aware that this problem was of public concern and that it provided a way of framing relatively new problems, a handful of power companies picked up on thinking like Saleeby's and fostered the distrust of smoke. In advertisements, Weil-McLain Boilers cautioned Chicagoans that their community sacrificed 37 percent of its light to smoke, a deep problem, the company contended, since doctors knew that pure air and sunshine were essential to health and welfare. Solvay Coke, which offered a cleaner-burning coal, put the potential loss at 50 percent and reminded readers of sunlight's worth as a disinfectant and "a powerful agent for the maintenance and promotion of health." In New York, the Consolidated Gas Company claimed that its industry's mission was to free Gotham from smoke, which one day would be "as unnecessary and as little tolerated as the open sewerage systems of earlier days." When clean fuel came, the city would return to "its rightful heritage of health and sunshine."[19]

For another group of health experts, however, none of the atmospheric observations really mattered. According to them, people in cities spent lots of time indoors, a disturbing reality since glass occluded all that was good and healthy in any sunshine. Beginning in the early 1920s, newspapers began telling their readers about this problem. By mid-decade, the occasional informational article turned into consistent, deeply worried hounding; one story in the *Los Angeles Times*, reprinted from the *Minneapolis Journal*, warned: "He who shelters himself behind a sheet of glass gets little more of this wonderful free medicine than he who hibernates beside a radiator in a dark room. Healthy sunlight is naked sunlight, and the only way to get it is to go out into it." In "Window Glass as Man's Enemy," the *Chicago Daily Tribune*'s health expert quoted a scientist who claimed that glass had cost thousands of lives.[20]

Eager for information about sunlight, Americans now had a composite picture of where it was not. The evidence came from diverse sources: researchers, government, and a popular and scientific press eager to sell copy and expand the public's awareness. Their conclusion: conditions were bad and getting worse. America's smokiest places were invariably northern, the same places that had the most to fear from other sunlight findings. That was where weather was cold and sunshine was weak, where children got little ultraviolet light, and adults spent all their days toiling behind glass. The problems of the north were mounting. Pollution, urbanization, overindustrialization, a sun that hugged the horizon in wintertime, and rotten weather combined to make a troubling environment.

The full magnitude of these problems only really became clear as medical researchers probed the diseases of darkness more closely and concluded that they had underestimated their prevalence and severity. Rickets was hard to diagnose. Certainly, few could miss the bowed legs of severe cases, but at the margins, there was less to indicate the deficiency.

Established in 1921, the Science Service Society was a clearinghouse for scientific news. With a broad educational mission, a close tie to industry, and a zealous passion for both, it produced informational material for multiple media, disseminated throughout America on a subscription basis. Among the Science Service's most common messengers was a simple photograph accompanied by a descriptive caption. One example presents an apparently robust infant with the caution that, though "this baby looks healthy . . . a physician's experienced

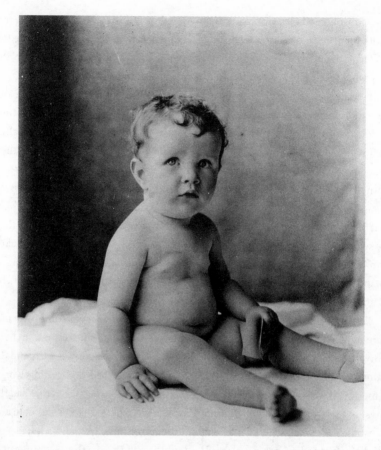

Figure 2.2. Mildly rachitic child, ca. 1920–40. Science Service Collection, Division of Medicine and Science, National Museum of American History, Smithsonian Institution.

eye would detect the bulging forehead and protuberant abdomen that indicate a mild degree of rickets, caused by insufficient vitamin D in the diet."[21]

Doctors could take many clinical paths to such a certain conclusion. An X-ray—or roentegram, as it was often called—seemed most definitive because it gave a picture of bone density. A blood serum test told doctors how much calcium a patient was retaining and therefore was also a good indicator of the body's vitamin D content. But by far the most common diagnostic tool was a simple clinical exam—the observations of a trained eye.

It is hard to tell, though, how good those trained eyes were, especially with rickets rates varying tremendously. In a 1917 *JAMA* article, ahead of its time, Alfred Hess advocated a prophylactic intervention to reduce alarmingly high rickets rates, especially among black children—the extra pigment of darker skin meant less vitamin D production.[22] In the article, he lamented the inability of clinicians to effectively parse the severity of cases. About fifteen years later, the researcher celebrated marginal improvements in the treatment of children. If he restricted positive diagnoses to only those showing symptoms in X-rays, 15–30 percent suffered, but a better estimate was 50 percent of white youngsters and three in four black boys and girls. Other assessments sounded more alarming tones, the *Delineator* claimed that 90 percent of all children aged six to eighteen months had rickets; in the *Literary Digest*, the statistic was three in four, and according to one 1924 *JAMA* report, 96 percent of all children eight months old were or had been rachitic and all black and Italian babies were (at the time, Italians were members of the set "dark skinned"). According to many authorities, this was the single most prevalent dietary deficiency in America. As notable as the deep concern over high rickets rates, though, was the variability in estimates of its prevalence; the *Los Angeles Times* commented, "Physicians of various schools have estimated that from 10 to 90 per cent of America's childhood is rickety to a greater or lesser extent." The article confessed that the reason for the variation in diagnosis was an "inability to accurately define" the condition.[23]

With rickets rates varying from one in ten to twenty-four in twenty-five and apparently healthy-looking boys and girls mildly rachitic, all of these pronouncements seem suspect. Historians of medicine have demonstrated that illness is in part a social phenomenon and that treatments often reflect the presuppositions of doctors and public health officials—hence, minorities, poor people, and women tend to receive different, generally inferior, and sometimes harmful care. There is another sense in which health is a social product. Rickets was real, but its diagnosis increased at a particular time in a particular place. Rickets rates rose because the condition was more common—a new environment had limited access to the most important source of vitamin D. But considering the vast variability in estimates of the condition's prevalence and the incredible ubiquity of diagnosis, it seems quite likely that rates reached their heights with help from doctors focused on finding the affliction.

Edgar Mayer boldly stated, "The loss of these effective rays from smoke pollution, glass windows and clothing must be stressed." The doctor's calls, no doubt, fell on sympathetic ears, and rickets became foremost among Saleeby's many "diseases of darkness." Rickets defined an actual condition, but doctors also diagnosed it because they lived in a time when, as the *Saturday Evening Post* reported, "natural sunlight is of first importance, and so far as it is concerned, our present civilization has run off the rails."[24] Rickets had become a way of talking about the troubling set of conditions that arose with modern, urban living. Its diagnosis, therefore, was sometimes an accurate medical assessment—often, it probably was not—but it was always social commentary.

The implications of a rachitic nation made medical critique all the more powerful. Doctors did not mince words when they cautioned about the consequences of the deficiency disease. Witness Charles Greeley Abbot's warning in the Smithsonian Institution publication *The Sun and the Welfare of Man*: "Rickets, as everyone knows, is a sort of lack of stamina, apt to invade the whole body of children." It causes in its little victims "a weak digestion, a poor appetite, emaciation, profuse night sweating, weakness of the limbs, tenderness of the bones, enlargements of the wrists and ends of the ribs, bow legs, curvature of the spine, misshapen head, contracted chest."[25] This sense that rickets was really a total-body disease helped shift thinking about the risks and benefits of sunlight.

In the 1870s, Downes and Blunt wondered if they had not found a clearly destructive force in the sun. Able to retard the growth of individual cells, light, it seemed reasonable to conclude, might be "devitalizing and injurious" to animals as well as bacteria. In time, that thinking evolved and sunlight became a qualified good, harmful to bacteria and relatively harmless to humanity. By the mid-1920s, even that hopeful picture was understated. Suddenly, it appeared, the sun's effect was nothing short of essential. The history of medicine has told a story about the first two-thirds of the twentieth century in which bacteriology featured heavily. After the discovery of the tuberculosis bacillus in the 1880s, scientists sought magic bullets—specific antibiotics for specific ailments.[26] Turn-of-the-century housing reformers told a different story—a critical environmental factor had the capacity to kill off all manner of infecting agents. A magic bomb rather than a magic bullet, this was a cure for any and all ailments. With regard to rickets, sunlight seemed to have even greater health benefits. It became part of a different kind of therapy, not just curative but constructive: bodies would be built strong with nature's help; this was a form of preventive medicine.

Rickets provided clues to one of the great curiosities of sunlight therapy. It was well known that some solar rays, especially ultraviolet rays, did not pass through media easily, and since the skin was a definite barrier, doctors struggled to figure out how a little ray of light could have large, systemic effects. According to internationally renowned British expert Sir Henry Gauvain, antibiotics were interesting, but "of greater importance and of fascinating interest is [ultra-violet light's] effect on deep-seated and inaccessible lesions to which by no conceivable means are they capable of penetrating." With vitamin D, the answer became clearer; opaque skin contained at least one intermediary. The superficiality of sun treatments did not matter because scientists found a mechanism whereby light hitting the skin affected an entire organism.[27]

Researchers worked hard to sort out the mechanism whereby light made health, but they never could come to consensus.[28] It was clear that the shortest wavelengths within the ultraviolet region penetrated least, but that was where the agreement stopped. The most common accounting held that the skin, and perhaps the blood in capillaries just below the surface of the body, absorbed sunlight, setting off a mysterious chain of biochemical reactions. Other doctors claimed that sunlight penetrated more deeply than commonly believed. Nearly all authorities agreed, however, that just what happened next was even more mysterious, and the promise of light therapy was breathtaking. If a ray hitting the surface of an organism could trigger the production of a just-discovered vitamin that traveled throughout the body, depositing lime in bones and stimulating untold biophysical processes, there was no reason to presume other substances did not act similarly. Perhaps vitamin D was good for more than science knew or sunlight irradiated additional powerful components of blood. Irradiated blood would then mix with other blood and pass on its positive effects. All of these entirely plausible mechanisms claimed for sunlight broad curative and restorative powers.[29]

From its start, sunshine theorizing came from international sources; in fact, early on, the United States was often a laggard. The first evidence that light killed bacteria originated in England, and it was to the Continent that the new sunlight-as-health-builder cadre looked. Well before Hess's German contemporaries did their work with rickets, and not long after Downes and Blunt's studies, a handful of innovators, including that open-air-school pioneer Auguste Rollier, were breaking new ground.

In the 1920s and 1930s, Paul De Kruif achieved considerable acclaim writing books and articles touting medical innovations. His stories of great healers were surely apocryphal, distorting the truth and turning doctors into demigods. That, however, did not limit his following. In Amsterdam, shortly before she and her family were seized, Anne Frank told of reading De Kruif's *Men against Death*, a book she held in high regard.[30]

Men against Death claimed that the field of heliotherapy had two great pioneers. The first, Dane Niels Ryberg Finsen, had an unfortunate childhood riddled by illness; fortunately for the world, as he grew older, he applied a deep scientific curiosity to these youthful experiences. According to De Kruif, Finsen's inquisitiveness led to observations that cats and bugs follow the sun around a room, moving to catch up with it as light and shadow shift throughout the day. His interest kindled, Finsen would not forget his lessons, in time devoting himself to the potential uses of light in medicine. Shortly before his premature death, the Dane received the 1903 Nobel Prize for pioneering work with artificial-light therapy. Some of his findings resonated with contemporary thinking popular in America, treating light and especially the chemically active rays as an antibiotic. Though his lamp therapy was far more technically sophisticated than the windows New York's housing reformers advocated, the principle was the same: light killed disease. Even before the Nobel committee recognized the Dane's studies, others had. By 1903, a handful of American serials were telling of work modeled on Finsen's in London, Paris, and Russia and reporting on the few American experts who had shown some interest.[31]

In some key respects, however, Finsen moved past the housing reformers and offered claims that resonated more fully with America's medical vanguard. By the doctor's own account, it was as much faith and observation as sound science that led him to advocate sun treatments. Having watched animals, including himself, grow stronger in sunshine, Finsen became confident that the relationship was more than mere coincidence. His faith was rewarded when he convinced other doctors that sunlight or specially designed high-ultraviolet-output electric lamps could have profound stimulative effects. He concluded from one set of experiments that solar rays "unlocked a dormant energy" trapped inside salamanders and tadpoles. This sense of sunlight's stimulative benefits was in line with the predominant American thinking in the '20s. It was also in line with De Kruif's other heliotherapy hero, Auguste Rollier.[32]

According to De Kruif, Rollier also took his inspiration from the animal kingdom. In the 1880s, the doctor was a pale, weak boy with a sick puppy and a sharp scalpel. Concerned about his pet, Rollier performed surgery to excise a tumor. Fortunately, he got all of the malignancy; unfortunately, the incision, dressed and hidden from light, failed to heal. It proved impossible to keep the animal from tearing off its bandages, but confounding the expectations of the budding physician, the wound, bathed in sunlight, soon got better. Later, as a surgeon frustrated that infection took the lives of too many patients, Rollier remained intrigued by the merits of sunlight. But it was not until he followed his tubercular sweetheart to the mountains that he became convinced of its healing properties. He reported that there he had found a medical Eden where his services were worth little because no one was ill. It was the crisp sunshine, Rollier

Figure 2.3. Auguste Rollier's Sunshine School, ca. 1920–40. Science Service Collection, Division of Medicine and Science, National Museum of American History, Smithsonian Institution.

concluded, that built mountain people so strong. This epiphany was the beginning of the doctor's vision. By the nineteen teens, he was treating children in his alpine hospital and school. For De Kruif, Rollier's lessons were clear; there was little that sunshine, stimulator and repairer, could not do: "It's our general resistance, the strength to stand off nobody yet knows how many sneaking deaths, that is raised to maximum by the energy zipping into our skins at a hundred and eighty-six-odd thousands of miles a second—from the sun."[33]

De Kruif believed that there was another considerable—though not quite equal—contributor to the movement, a slightly lesser luminary: "[Rollier,] the chamois hunter, city-hater, sun-dreamer, took pure fire from [Oskar] Bernhard, who'd taken the fire from Finsen." Others disagreed with the lay medical expert, promoting instead Bernhard's St. Moritz clinic as a pioneering institution on par with any other. According to De Kruif, the medical middleman was fond of the saying, "Where the sun is the doctor ain't."[34]

By the time De Kruif wrote in the late 1920s, the United States had caught up in its heliotherapeutic enthusiasm. Benjamin Goldberg, professor of medicine at the University of Illinois and director of Chicago's Municipal Tuberculosis Organization, wrote of his own experiences with animals. Light, he said, was responsible for all animate life, a healer and a comforter. It was little wonder, then, that the lower species were "dawn worshipers":

> Anyone who spends a night in the woods, in the spring or summer, is aroused at dawn by the universal bird chorus of jubilation. Cock-crow, the reaction of the cock to the glory of the coming day, has become synonymous with the dawn. In the jungle the apes, in the mystic moment as the new day is born, beat their breasts in frenzy and shout an inarticulate hymn to Apollo, the sun-god.

Authors talked about the blossoming of young bodies after a dark winter or of sunlight's mysterious ability to build strong people: "Whenever body energy is deficient, heliotherapy finds its indications. Bacteria, or their toxins, lack of vitamins, calcium deficiency and other agents of enervation acting upon the human body may be regarded as the causes for general constitutional defects and secondary local manifestations."[35]

With sunlight authorities enthusiastic about their new treatment, compelling evidence of its potency, and little to indicate its limits, the laundry list of ailments it could cure grew long, and consensus frayed. Temple University's director of the Department of Physical Medicine, Frank Hammond Krusen, offered the following chapters in *Light Therapy*:

Chapter 9—Ultraviolet Radiation in Diseases of the Alimentary Tract
Chapter 10—Ultraviolet Radiation in Diseases of the Circulatory System
Chapter 11—Ultraviolet Radiation in Diseases of the Respiratory System
Chapter 12—Ultraviolet Radiation in Diseases of the Nervous System
Chapter 13—Ultraviolet Radiation in Diseases of the Bones, Joints and Muscles
Chapter 14—Ultraviolet Radiation in Skin Diseases
Chapter 15—Ultraviolet Radiation in Genitourinary and Gynecological Diseases
Chapter 16—Ultraviolet Radiation in Diseases of the Eye, Ear, Nose, Throat and Mouth
Chapter 17—Ultraviolet Radiation in Systemic and Miscellaneous Diseases

The Philadelphia physician was not alone in singing the praises of sunlight. Others divided their works differently, but generally, these medical tracts proclaimed broad biological powers for sunshine and ultraviolet light. They both celebrated the reach of scientific understanding and confessed that experts still had much to learn. Incomplete knowledge, however, was only a small problem. In the introduction to their book *Light and Health*, which proclaimed a huge role for light in the promotion of wellness, lighting engineer Matthew Luckiesh and his colleague A. J. Pacini argued that solar radiation had been a powerful evolutionary force. That being the case, a world without it was inconceivable: "To assume that it is no longer of importance to the health of human races is as unreasonable as to assume that oxygen is no longer essential to the respiratory process." With such a bold pronouncement, it was unsurprising that the authors ascribed wide-ranging curative powers to light.[36]

The uses of sunlight had grown legion. Rickets remained the unquestioned king of the diseases of darkness; it was the only one that sunlight treated fully and perfectly. Tuberculosis, especially the extrapulmonary variety—tubercular lesions can, in fact, lodge in lots of places—was next, but no less notable was the gross proliferation of lesser conditions sunlight could treat. If a young man went to school at Cornell, he could expect prophylactic sunlight therapy for the common cold. G. H. Maughan and Dr. Dean F. Smiley conducted a series of studies at the Ivy in the late 1920s to test the effectiveness of light treatments on cold-susceptible college students. Articles in the *Washington Post*, the *Chicago Daily Tribune*, and the *New York Times*, told of the Ithaca instructors' impressive results: a 50 percent reduction in winter colds. After further experimentation and a couple of years of success, Dr. Smiley announced that the school was expanding its treatment facilities. In 1930, it would add a student-accessible sunlamp, available for a small fee, to its ultraviolet solarium.[37]

Indeed, it appeared, sunlight's potent stimulative and antibacterial cocktail could do just about anything. Authors wrote books about its role in dentistry: increasing calcium in teeth, and destroying bacteria infecting the mouth. In *Good Housekeeping*, Walter Eddy claimed that an 8,000-dentist study had found tooth-decay rates higher in the north as a result of less sunlight than down south. In the mid-1920s, two Chicago hair-loss clinics battled over which offered the most modern therapy. The ads were small but ubiquitous—their claims grand. They promised patients a specialized form of treatment that used lamps to create ultraviolet light: "Great inventions are astounding the world! The success attending the use of ultra-violet rays in growing hair on bald heads is no more wonderful than scores of other new-age marvels." The reasoning behind this particular treatment was simple. Arnold Lorand's book *The Ultra-Violet Rays: Their Action on Internal and Nervous Diseases and Use in Preventing Loss of Color and Falling of the Hair* explained why people went bald. It was not natural, and it was not necessary. A strong medical intervention could stem the tide of age-ing. There were two main causes of baldness: infection and poor blood flow. Ei-ther of these problems left the follicles poorly nourished and hair liable to fall out. Lorand's treatment took care of both; a short ultraviolet blast to the scalp would kill troublesome bacteria and feed the follicles by stimulating blood flow to surface capillaries, increasing hormonal secretions, accelerating metabolism, and exciting the nervous system. Hair-growth experts were clear: sunlight was both killer and nourisher.[38]

It would be a mistake, however, to simply say that every doctor believed ul-traviolet radiation would cure baldness. Medical textbooks demonstrated the divisiveness regarding sunlight therapy. Some authors thought that they had found a broad and important curative, but many saw a more constrained role for sunlight as an important germicidal agent, able to kill bacteria or mold. Still others thought that its role as bactericidal and health agent was even more cir-cumscribed. They did not doubt that sunlight would cure rickets, and they con-sidered it a disease killer, but only of tuberculosis. And a final group thought that even when it came to tuberculosis, sunlight was an imperfect treatment. In fact, they said, it was only certain kinds of tuberculosis, especially tuberculo-sis of the skin, that responded well to sunlight. This lack of agreement does not mean the doctors who taught about light's widespread merits in medical col-leges or wrote up research that celebrated the sun were marginal or insincere. Whatever the caveats, light therapy did seem to hold great promise.[39]

Tenement reformers had spoken out in favor of better-lit houses, but by the late 1920s, sunlight was not sunlight pure and simple. Scientists, who had un-covered the mysteries of ultraviolet radiation, staked out a set of claims demon-strating that to be fully worthwhile the solar spectrum had to come unfiltered.[40] Glassmakers had more to lose from the new discoveries than anyone. Their win-dow products, once the hope for a bright new urban future, had suddenly be-come an integral part of the problem. In the journal Hygeia, a publication of the AMA aimed at a popular audience, Charles Sheard made the point: "For today we are living almost entirely in glass houses behind glass windows; yea, even in our shut-in and glassed-in automobiles. No cheeks of tan; no sunshine on the face. Surely civilization gets its price for what we wrench from it."[41]

But not all glass, it turned out, blocked ultraviolet light or blocked it simi-larly. Innovation might save glassmakers from the abyss. Glass businesses of-fered readers scientific evidence that fed sunlight fears and bred faith in new products. Combining, science, social critique, and salesmanship, they told of humanity's rightful place in nature, of the danger of cities, and of the value of innovative glazing. In 1927, Vita Glass, a new and promising product invented by Trinity College fellow F. E. Lamplough, offered American consumers health and wellbeing. In ads, companies pushing Lamplough's ultraviolet-porous glass and similar discoveries articulated many sunshine worries and offered simple fixes; meanwhile, science provided its powerful sanction.

Vita Glass assumed an audience already well acquainted with the nature of light. Many of its ads feature a prism, dividing light into its components: visible, ultraviolet, and infrared. All three beams shine onto two panes of glass, one ordinary, impermeable to light's shortwave healer, the other vitalizing. In some of these ads, two rooms are shown side by side. They look the same except, in one, a hand reaches down and pulls a clear shade from over a window. That room is brighter, and the caption reads, "Vita Glass like an unseen hand raising an unhealthy film, lets them pass indoors." In "Do You Still Live in a Cave?" the company tells the complete story of human civilization in one page. History began when men emerged from caves thousands of years ago and decided to set up camp on the sides of "sun-drenched" cliffs. But, recently, the advertisement warns, man has returned to "caves with window glass," caves that exclude the sun, that make him weak, that prevent him from remaining "the creature of light that he is, rather than the soggy thing that crawled out of the dark when the world was young."[42]

Newspaper advertisements regularly presented readers with dire warnings. In the ad "You Spend Your Day among Thieves," readers are told of the robbers they meet every day who "filch something more precious than money, they steal health":

> Reach out and touch the pane of window-glass at your side, and you have laid hands on one of the highwaymen. You met a pal of his at breakfast, and sat by one on the train. You work all day behind another pane of common window-glass, whose deceptive surface deprives you of the cream of the sunshine.

In other ads, the message was similar, but Vita Glass turned the metaphor upside down. Subject to an "unwittingly cruel" new reality, a businessman sits at his desk, dressed in prison stripes; he is one of the many "prisoners of glass!" He and his fellows have been "deprived of the health-giving rays of the sun, as no previous generation ever was. The sunlight may stream into our houses and offices through closed windows, yet it is utterly deficient in healthful properties."

Vita Glass pushed a new enemy to the fore of the sunlight debate: work. Increasingly indoors, laboring behind windows, Americans endured a set of barely sustainable conditions.[43]

Vita Glass—in some ads, translated as "glass of life"—offered its buyers hope and told of dangers repeated frequently in the medical literature: harmful winters, diminished resistance to infection, a feeble body, and undernourished blood. Installing the Vita Glass product could help prevent that "rundown winter feeling." One pitch asked "Do you believe in sunlight?" another told the story of "what winter means to New Yorkers" in ten words: "Chill days ... indoor living ... lost vitality ... chronic colds ... minor illness." The reason for those ten words was two: "Sunlight Starvation."[44]

But Vita Glass was not alone in its claims. Fused quartz was an older glass innovation that offered healthful sunlight; it did not pilfer ultraviolet rays. Quartz-Lite—a brand that employed this good substitute—repeatedly used scientists to convey its message. Ads in the *New York Times* and the *Chicago Daily Tribune* referenced scientific proof that regular window glass left rooms sunlight poor, rendering children rachitic and adults vitamin deficient. This lamentable situation, according to Quartz-Lite and its bevy of experts, kept the sun's rays, "Nature's Invisible Tonic for 4 O'clock Fatigue," outside: "Those invisible messengers of health ... are tapping on our windows for admittance."[45]

In pamphlets probably intended for salesmen, Lustra-glass, a product of Pittsburgh's American Plate Glass Company, assumed a savvy readership. The company was honest, explaining that sunlight did not provide light below 300 mμ and admitting that at the shortest region of the sun's spectrum, Lustra-glass did not transmit much, only 2.3 percent of waves at 302 mμ. But for the slightly longer ultraviolet rays, the product was a boon to health, permitting passage of 71 percent at 334 mμ. The relatively poor permeability to lower wavelengths may have led to the confession in another pamphlet that Lustra-glass was not represented as a "scientific health glass"; nevertheless, it did convey "a substantial amount of the therapeutically effective ultra-violet rays of sunlight." This linguistic hair splitting, however, seemed unimportant alongside evidence of the product's benefits for sun-starved chickens and wilting spinach.[46]

These were only a few of the products sold as window substitutes to a light-hungry public. In 1929, an article in *Scientific American* asked, "Ultra-Violet Transmitting Glass—Has It Made Good?" The article responded to a sea of inquiries from its readership, concluding that innovations like Quartz-Lite were "doubtless destined to have a permanent future for health purposes, but that

[their] usefulness is likely to be more limited than some of [their] more enthusiastic proponents think." The AMA positioned itself as the clearinghouse for
all information on the subject of glass innovations. In 1927, it passed on the results of Vita Glass's tests on its own product, which unsurprisingly found a most
promising innovation. A subsequent article on glass substitutes offered findings
about other products: Flexoglass, Celoglass, Corning's special (Corex) glass,
and Cathedral's translucent option, good for maintaining privacy.[47]

One problem with all these substitutes, however, was that they degraded over
time; solarization, the natural process by which old, ultraviolet-porous glass became opaque, could reduce health benefits. One 1929 article claimed that solarization was a minimal problem. At the United States Bureau of Standards, W. W.
Coblentz disagreed; and most people sided with him. The federal government
assumed a supervisory role here and in that capacity touted some products over
others. Coblentz, along with his frequent collaborator R. Stair, studied the solarization problem extensively. In 1928, they offered their findings, which tested
Celoglass, Vita Glass, Helioglass, Corex A and D, quartz glass, and Locke Glass,
concluding that nearly all had their own particular and significant transmission
problems—with quartz glass the near-perfect exception.[48]

The researchers found that there were far more products than they could examine, including London's Holviglass, Paris's Renovic, and Nuremburg's Pret-
let. The robust market, international in scope, led Coblentz and Stair to conclude, "This country seems to be a rich field for exploitation and naturally the
tendency will be to cheapen the manufacture at the risk of possible sacrifice in
the transmissive qualities of the glass." With such potential dangers, Coblentz
sought tighter supervision before the problem grew out of control. In 1930, he
decided it was time to change the stringency requirements for "ultraviolet transmitting glass." The new focus did not worry about shortwave ultraviolet rays,
absent in significant quantities in natural light, but would require permeability
to the healing rays the sun did transmit.[49]

With products like Lamplough's, there was hope for glassmakers, but they
were not the only innovators looking to capitalize on sunlight findings. The argument other inventors made was simple, and they had both reason and science
behind them: while a Lustra-glass bungalow placed in a beautiful alpine meadow
might have brought health to those who lived within, for most homeowners—
and nearly all urban renters—the fundamental problem with glass could not be
so easily resolved. Windows were good only where powerful sun shined.

Tenement reformers acknowledged this commonsense problem when they talked about overstuffed lots that shaded neighbors. In the 1920s, however, concerns about windows reached a new level, and Americans witnessed a proliferation of lamps looking to pick up where the sun left off. These products, made by General Electric, Westinghouse, the Hanovia Chemical Manufacturing Company, Sunray, and others, generally purported to offer a light that simulated sunshine in its full spectrum. Some claimed even greater specificity. Hanovia, for instance, picked up on the association of altitude with health and contended that although there were other products on the market only its Alpine sunlamp gave "a scientifically correct and safe concentration of ultra-violet rays."[50]

There were two primary types of lamps. One used two carbon electrodes. A charge crossing the electrodes heated them and caused them to emit a spectrum approaching sunlight's. The mercury vapor lamp was less sunlike, but more powerful. Its bulb contained electrodes and a pool of mercury. A charge heated the mercury and vaporized it. The combined radiation from the electrodes and the excited gas caused the lamp to emit a potent light with the ability to tan. The more primitive—and cheap—carbon arc lamps often did not use bulbs at all. Mercury lamp makers, however, needed something to contain the liquid, and when in use, the gas. For that reason, they owed glassmakers thanks; without the invention of fused-quartz lightbulbs, mercury sunlamps would have done little for health.

In the 1920s and early 1930s, it would have taken an abacus to count the bathing-suit-clad youngsters in sunlamp ads. They sat on their mothers' laps, played with trucks, hugged puppies, and donned dark glasses. In other ads, glamorous-looking women relaxed or put on makeup beneath lamps, and shirtless men lounged in front of their own suns. Some of these ads preyed upon the same fearful associations that Vita Glass did. One pitch by the Commonwealth Edison Electric Shops featured "the sunless cavern" most people endured in the workplace and another, by GE, lamented "modern conditions." But a Commonwealth ad, published in the *Chicago Daily Tribune*, also made a move not available to glassmakers when it reported that, in December 1929, the sun shined for only 25 percent of daylight hours and that the following month was less than half as sunny as was February. General Electric reported that the winter sun shined even less, on average only twenty-three days during the entire season. This was precisely the point Eveready's Sunshine Map had made. Glass could help only with nature's cooperation. Electric light did not suffer such a limitation.[51]

A consortium of New York electric companies showed the failure of winter sunlight with two small graphics—one showing the summer sun shining straight down on the city with the caption "direct rays—long days—clear skies—full ultra-violet effectiveness," the other showing winter sun's "slanting rays—short days—cloudy skies—1/8 ultra-violet effectiveness." This sun was no friend; it was "an old swindler," a "fraud." Indeed, if glassmakers taught Americans to fear life indoors, the sunlamp taught them to fear winter. Eveready celebrated those who went out and got "stimulating" January exercise but cautioned them that they were not capturing "all the body building rays of the mid-summer days." According to General Electric, "It has been known for ages that sunlight is a great help in maintaining health and resistance to sickness . . . in developing stamina and under proper conditions, in generating buoyant conditions." That was why people felt better and were happier and more carefree in summer. Eveready's 1929 ads "Why Risk Your Health in Winter?" and "Why Do We Dread the Cold, Dreary Days of Winter" claimed that its product was the way to keep summer health all winter long, that it prevented colds, built up resistance, and stimulated glands. Eveready begged readers, "See for yourself how the healthful light from these lamps fills your body with sun-heightened health and vigor . . . just as if you had been basking under a June sun! For their light supplies vital rays you fail to receive from winter sunshine."[52]

Some advertisers claimed almost supernatural effects for their sunlamps. Readers were called upon to bring the sun into their homes; its properties were magic, its power as tonic profound. Far more commonly, however, sunlamps were described as gifts from science. According to its makers, in the "so-long" struggle of scientific advancement, the L&H Super-Sun was the first product to replicate sunlight in all its glory: "Truly, man has created a sun—captured within the compact confines of a simple cabinet whose doors are the doors to health."[53]

The ties binding medicine, technology, and pure scientific research were strong. Sunshine was a newly (re)discovered health agent and one that pitchmen enthusiastically sold. Often, ads mentioned doctors who supported their products and scientific evidence that "proved" their efficacy. Around 1940, Eveready began printing a catalog for its salesmen, offering them all the evidence they needed. Binders were divided into five color-coded sections, "General," "Tuberculosis," "Pediatrics," "Dermatology," and "Miscellaneous." Each contained abstracts and citations to scientific articles that confirmed sunlamp results. Pagination began anew with each category, and Eveready promised vendors regular supplements with newly published reports. The binder married business and science to invention in an effort to sell a product to the public.[54]

Historians have written on the role that artificial light played in transforming urban life at the turn of the century, but they have told only part of the story. Scholars trace the emergence of powerful new technologies and their profound effects: pitch-black streets were gone; urban boulevards became places to enjoy after dark; night wanderers, criminal and lawful alike, struggled to escape supervision; bright, beautiful lighting displays astounded; and increasingly, affordable lamps permitted homes and workplaces to remain active twenty-four hours a day. But little of this impressive scholarship has looked into fears of an emergent lighting danger: according to concerned citizens, the same inventions that made dark places light were beginning to lure humans from their natural place in the sun into urban "caves." The newly populated environments of the modern city were trouble precisely because all light was not the same; real sunshine was what people needed.[55]

Five billion years after its creation, the sun, according to Vita Glass, had finally fulfilled its potential. In an ad that put rickets rates in northern cities at 97 percent, the company asked of a featured child, "Is This Little Fellow Worth $25?" That was what it cost to bring bright, invigorating sunshine to a sun-starved body; that was the cost of one Vita Glass window. Lamp makers advertised that they could supply sunlight for a variety of budgets: $12.50, $26.50, or $69.50. They said that lamps cost only one cent to operate for twelve minutes or three cents an hour.[56]

In a *JAMA* article, A. H. Pfund suggested a more democratic approach that would make sunlight available to all. Aware that the poor needed to avoid debility and concerned that they might be most susceptible to illness, Pfund proposed a window solution within the reach of any American. It was a wood frame with chicken wire across the span and cellophane pulled tight on top. Do-it-yourselfers could remove and reinstall this low-cost "window," which could last up to a year. But even with this bargain, it was clear that light now generally came at a cost and that optimal conditions would probably remain beyond the reach of poor Americans. That reality could only have occurred once sunshine was a scarce resource, which was precisely what it had become.[57]

Unfortunately for the companies touting light enhancers and substitutes, the public was not especially eager to pay up, and many products struggled to capture market share. In 1932, General Electric promoted its S-1 sunlamp aggressively, but the public's response was tepid: just under 55,000 bulbs left stores in the first year, far less than 1 percent of the lightbulb market. Sales were poor in part because consumers were unlikely to use these bulbs in more than one unit or for more than a small fraction of the day, facts General Electric must have known; still, the company could not have been happy, and it would see sales drop further in the next year. The S-2 lamp faired even worse.[58]

Vita Glass's failure was no less spectacular. In 1928, the British company worked hard to succeed in America, with half of its inventory of 80,000 feet of glass per week headed across the Atlantic. Its star faded in the face of competition from rivals, limited public enthusiasm, and concerns about solarization. Other companies may have been more successful, but health glass clearly fell short of businessmen's hopes. By the mid 1930s, advertisements for their products had declined precipitously.[59]

It is hard to know what to make of these failures to infiltrate the lighting and glassmaking markets. None of these products were major commercial successes, and few made it into all that many homes, but there are other measures of importance to be considered. John Sadar argues that Vita Glass gained "mindshare" even if it failed to capture market share; as I discuss in more detail below, it was on the tips of tongues in the public housing and public health debates. The same can be argued about other light products: doctors put health lamps to use in their offices and directors put them in Oscar-winning movies. Moreover, this enthusiasm was part of a larger movement to secure sunlight and suggests a developing understanding of the environment.[60]

According to Vita Glass marketing, Nero once paid the equivalent of $100,000 for "the marvel of his age," a piece of glass. The company's new wonder was therefore a bargain by comparison, cheaper and better for consumers. But Nero's enthusiasm also betrayed a primitive notion of light. According to Vita Glass, the emperor did not simply overpay; he paid for fake sunlight. Roderick Nash's *Wilderness and the American Mind* (1967) helps to explain the logic of this argument. Nash put ideas about nature in historical context—at different times nature connoted danger, magic, serenity, or opportunity. With sunlight we can take this historical phenomenon a step further, for what qualified as "natural" about sunlight also changed.[61] In 1915, planners hoped for a future of sunlit rooms within well-organized cities. By 1925, with the twin discoveries of rickets and the fragility of ultraviolet light, that promise lost its appeal. People had long understood that sunshine was a composite; suddenly, light streaming through a clear window was unnatural—bereft of a key component. This was updated nature: pure, scarce, and potentially valuable.

With the public's changed sense of what qualified as sunlight, companies sold their own modern versions of nature, blurring boundaries between synthetic and real. In Eveready's ad "Growing Children Need Sunshine in Winter Too," a small parade of five lightly clad children play in wagons and push carriages. A companion image shows three children sitting with a ball and some

Figure 2.4. "The UVIARC Sun Lamp for Healthful Sun Baths," 1929. General Electric Vapor Lamp Co., Trade Literature, National Museum of American History, Smithsonian Institution Libraries.

blocks as their mother focuses a lamp on them. In the ad, natural and artificial have become kin. A promotional pamphlet for the UVIARC Sun Lamp, a GE product, made them more than relatives. On the cover (the same picture appeared inside another booklet), two children sit on the floor of a room. They wear bathing suits and play in sand. In front of them a sunlamp lights their play space. Behind, a chalkboard shows a drawing of a sun, a sailboat, and an ocean. Another pitch did not limit its subject to children, spelling out what these promotionals represented in pictures and issuing consumers a set of instructions:

> Take your sun bath during the morning shave. Clothe the children in the popular sun suits and let them play under the lamps each day with their toys.

During the afternoon rest or reading period partially disrobe for your daily ration of sunshine. Take the sun bath in a reclining chair or on a bed before retiring at night.[62]

These lamps and glass products reveal a lot about how businessmen, advertisers, and inventors saw their world and how people increasingly understood technology. Historian Roland Marchand has written of the modernist advertising that celebrated innovation. The story he tells in *Advertising the American Dream*, however, is not simple. For Marchand, "the parable of Civilization Redeemed" joined anxiety over modernity and a celebration of the past to a confidence that progress and technology could return the best qualities of nature. Lighting engineers expressed similar fears and hopes when they talked about creeping darkness but sought innovations that would make the world better than ever before.[63]

This was not a time dominated by retrograde traditionalists or forward-thinking modernizers, people who blindly embraced nature or unabashedly celebrated innovation. As the twentieth century dawned, with the country in the midst of rapid urbanization and industrialization, it would only make sense that people understood their futures—brighter, lit by lamps, or darker, dominated by smoky cities and windowless rooms—more complexly. Undoubtedly, some hid from the future and others charged toward it. But the lamp company that wrote the following testimonial for its product is tougher to sort out:

Recent research has proven beyond doubt that the ultra violet region is of the greatest importance to human welfare from the physical standpoint. Without this type of energy reaching the body direct, human beings, and most animals suffer a type of light starvation which leads to general poor health, anemia, rickets, susceptibility to colds, etc.

In times past when man lived largely out of doors, these conditions were not prevalent, due to long exposure to sunlight and unpolluted skylight. With the evolution of our modern civilization man came indoors and literally shielded himself from the sun. Our modern system of housing and clothing at once became a perfect insulator to sunshine, and all its benefits.

The fix for folks enduring these conditions was a sunlamp. Ironically, the champions of progress struggled to explain the world they were ushering in. They sought better lives and lifestyles; they looked forward, but their promise was

that, by embracing the future, people could recapture what they had left behind. Technology would restore natural bright light and good health to every
American.[64]

But the lamp maker and glazier were only suppliers. They themselves were not
the ones who put sunlight, natural or manmade, to use. In institutions across
the country, animals, from humans to chimps and chickens, were being lit up.
The techniques used to make the sick healthy and the rest vigorous varied, but
all experts counseled that they had to be precise and based on sound, scientific
principles.

3

Sun Cures

Inspired by its author's experiences as an ambulance driver for the Italian army in World War I, Ernest Hemingway's *Farewell to Arms* starts with love but moves quickly to the horrors of war. Sitting in a hospital after a mortar shell wounds him badly, the young protagonist receives a diagnosis from a triumvirate of semicompetent doctors. Surgery is a must, but only after six months of healing. Aghast at the prospect of such a long period idle, he calls the house doctor back for more counsel and is told that he will be able to use crutches but only after the wounds have been suitably exposed to the sun. This initial period of "sun cure" followed by light exercise will ready him for the trauma of surgery. Ultimately, Hemingway's character finds a better doctor willing to operate immediately; the recovery includes "baking in a box of mirrors with light rays." At a time when America's heliotherapeutic regimens were stuck in a prolonged infancy, its son was off in Europe embracing them.[1]

Hemingway's experience is important because it makes clear that the sunlight story is not just American or only one of wavelength, pollution, or clouds. It is also the story of hosts of people

around the world crafting and experiencing treatment. As evidence of heliotherapy's merits spread, so did its use. It was a nationwide phenomenon with
hotspots in Louisiana, Indiana, Pennsylvania, and Arizona. However, doctri-
nal conflict was common. In the 1920s, medicine was a fast-changing profession, having recently emerged from a time when training was minimal, treatment
haphazard, and oversight nearly nonexistent. Still out to prove their legitimacy,
practitioners were profoundly concerned with rigorous, scientific, standardized
treatment. Doctors today would question their practices, scientists today would
critique their research, and the public today would probably walk away from
their care, but that does not change the fact that these were trained professionals.
In 1925, *JAMA* editor Morris Fishbein wrote of medical quacks; heliotherapy
was nowhere included. The fractiousness of medical care that Hemingway experienced was common, but heliotherapy was by no means a fringe treatment.[2]

With lots of contradictory claims regarding light treatment, this part of
sunlight's history is messy. Sanitarium administrators and doctors became the
most prominent voices in the debate, but theirs were not the only ones; other
people—including public health officials, veterinarians, and educators—
claimed a role in devising and promoting treatments. In little time, negotiating
the borders of sound science and naturalistic approaches, more wide-ranging
cures emerged, with elaborate apparatuses and wild promises. The medical
profession struggled to control the new treatments largely because it could
not close ranks around a set of principles. With a sun-starved public, sunshine
began going where before it had never been.

It made sense that people in places with skyscrapers and thick pollution worried about lost light, but heliotherapeutic treatments were hardly limited to big
cities or the industrial Midwest or Northeast. Articles on the subject appeared
in *Minnesota Medicine, Northwest Medicine, Virginia Medical Monthly,* and the
West Virginia Medical Journal. One session of the AMA's annual meeting on
pharmacology featured the paper "The Use and Abuse of Heliotherapy in Tuberculosis," which drew commentators from El Paso, Denver, and Phoenix.[3]

With little consensus and a still-young understanding of light, theories about
how to dispense the sun abounded. Nevertheless, doctors were little interested
in ad hoc, willy-nilly treatment; this was science, and a type of literature review,
synthetic in nature, exhaustively researched, and extensively referenced became
common. To be modern and medical, therapy needed well-considered dosages
and a prescription for treatment. As it turned out, such precision was most difficult to achieve.[4]

Researchers knew that the body did not use all wavelengths similarly, and that made light hard to prescribe. Most practitioners divided the ultraviolet spectrum, roughly waves below 400 mμ, into far, middle, and near (some had only two categories). For the body, the most important solar waves were the ones between 290 mμ and 314 mμ, the lengths best able to make vitamin D and most effective at tanning. Scientists and doctors counseled that the shortest waves in sunlight—ranging down to 265 mμ under ideal conditions, but really only in practical quantities down to 290–300 mμ—were the best antibiotics. Even shorter waves were more potent, but they did not come with sunshine. Thus a thin band, wavelengths 290–314 mμ, was what piqued medical interest. Doctors contended that the problem with winter sunlight was simple: light from close to the horizon traveled through extra atmosphere, which occluded that 290–314 mμ band. Though authorities inaccurately ascribed powerful bactericidal properties to sunlight, misjudged the dangers associated with all ultraviolet light, and underestimated the development of young bones during winter, the general principles doctors established were correct, and all of their findings—those that future researchers would validate and those they would not—informed treatment.[5]

Clinicians concluded that there was little sense in prescribing two hours in front of a glass window that permitted the passage of no ultraviolet radiation, but it was not much better to suggest two hours in a specially glazed solarium if its glass did not have the requisite transmission properties. Lustra-glass, for instance, with its exceptionally high permeability at 334 mμ did little for most wellness seekers. In contrast, Edwin T. Wyman's team of researchers reported, Vita Glass transmitted 50–60 percent of all rays emitted by a Cooper-Hewitt lamp in the region 260–300 mμ, and Corex roughly 92 percent. That information had only limited value to the precise practitioner, however, since solarization meant that two-month-old glass was vastly different from new glass.[6]

Clinicians also struggled to determine how to prescribe treatment with lamps. Different types of units emitted different spectra. Carbon and mercury lamps, of course, were vastly different. But even within each category, there was considerable variability, as companies produced unique burners with different types of electrodes: a Hanovia lamp was different from one made by General Electric. The National Carbon Company produced B, C, D, E, K, U, and W carbons, all emitting different spectra. Much as was the case with window glass, lamps' transmission patterns changed over time: their bulbs, if they used them, could become solarized, and the carbons they employed degraded. No two moments in front of a lamp had the same effects.[7]

The unfortunate reality that people responded to the sun differently compounded these problems. On average, brunettes tanned much better than blonds—extra pigment allowed longer exposure without bad burns—and increased photosensitivity, either due to pigment or other skin idiosyncrasies or from diet, could cause similar inconsistencies. One article in the *Archives of Pediatrics* summarized the challenges of treatment succinctly: "It is not possible to prescribe an exact dosage as various burners vary in their intensity over a fairly wide range and the sensitivity of the individual patient differs in each case."[8]

The often-mysterious effects of light only made standards for therapy more elusive. As one author put it, "The results of exposure to the radiation are often of an intangible, invisible, physiological nature which renders them difficult to evaluate." Still, this uncertainty was no contraindication for treatment. Rather than retreating in the face of an inexact science and indefinite effects, doctors sought answers as best they could and cherry-picked evidence. They achieved a half-understanding and employed clinicians' experience to shape a treatment rigid in its outline but flexible in its application.[9]

In order for therapy to be scientifically rigorous, doctors needed an understanding of what it was they were prescribing, and that required a way of measuring the amount of biologically active light in natural or artificial sunlight. Some shined rays on a liquid: the chemically active wavelengths, if present, catalyzed a reaction and turned the solution a different color. Others substituted a photoelectric plate: high-energy wavelengths kicked electrons off the plate, which with amplification gave a reading of the light's power. But the body was not a photoelectric plate, and it was not a solution. The wavelengths that affected them might be different from those that improved health. With carbon arc and mercury vapor lamps emitting in their own unique patterns (often at wavelengths well outside of the solar spectrum's distinctive footprint), doctors had no way of knowing what mixture of wavelengths they should measure.[10]

The preferred alternative to such devices was not technical at all. Doctors wanted to separate the light that had biochemical effects from light that did not, so many decided to use humans as their measuring standard. Their sunlight regimens had as a goal in prescription "minimum perceptible erythema" (MPE). However, there was less agreement on what, precisely, erythema meant. For most physicians, it was the initial reddening of the skin caused by light; for others, many of whom correctly noted that ultraviolet radiation did not cause that initial coloration, it was roughly synonymous with tanning, the darkening of skin about a day after treatment; and for a few, it was an intermediate effect about four to six hours after exposure. A dose was the shortest period of time necessary to produce this coloring. Most doctors were careful to point out that neither erythema nor a tan actually made a patient well. Rather, they were indicators that sunlight was working.[11]

Though dosage varied among patients, the specific method of practice was rigid. Lamps had to shine straight overhead because any angle limited light's ability to penetrate the surface layers of the skin and therefore decreased its effectiveness. The distance from source to subject required careful calibration, since light waves halved in power for each doubling of distance. Doctors offered formal but flexible guidelines of how best to treat each patient. Many who did not use the MPE system—and some who did—started with a test strip, little different from a photographer's. Patients would expose a relatively untanned part of their body (often the underside of the arm) to light. By covering different parts of that surface for different lengths of time, a doctor could gauge photosensitivity and determine a personalized treatment regimen.[12]

Auguste Rollier was a pioneer in this work. He developed test strips, and his program of progressive desensitization, also called zoning, intended to prepare patients for treatment without endangering their welfare. Patients began a regimen covered up, only gradually baring their skin to sunlight. In natural heliotherapy, average exposures began at five minutes once or twice daily around the feet and ankles. In subsequent sunbaths, doctors increased the length of treatment for previously exposed body parts and added other areas over a period of days, first legs, then arms, then back, and finally the chest. In many copycat systems, the standard period of treatment was different, and other practitioners deemphasized zoning, opting instead to focus entirely on a graduated duration of exposure. Therapy with lamps high in low-wavelength ultraviolet radiation required a further adjustment: the power of their rays necessitated increased precision and shorter treatments, but the general principles remained the same; no part of the body should receive sunlight until it was ready, and all patients had their own needs.[13]

The twofold merits of sunlight meant a two-pronged approach. Systemic—or general—therapy subjected the entire body to treatment. The thinking was simple: if sunlight built up resistance to infection, it should shine on all parts of patients. But light could also kill disease where it started, so in addition to generalized treatment, practitioners advocated local radiations.[14] Indeed, ultraviolet therapy started as a treatment for surgical tuberculosis, radiating the site of an infection, and it never fully shook that heritage. Badly burning diseased skin, doctors believed, could work wonders in ways that general treatments simply could not.[15]

Authorities claimed that this medicine had to be well considered and that serious, sophisticated facilities for its dispensation were no less important. Predictably, hospitals were the most common—but by no means the only—type of heliotherapy facility. These places did not resemble twenty-first-century institutions. They ranged in kind from big buildings in inner cities to large campuses (more like present-day camps) that allowed struggling youngsters to recuperate away from unhealthful urban settings. Historian Gregg Mitman has suggested that the rise of bacteriology resulted in a decline of enthusiasm for natural curatives for tuberculosis. In the nineteenth century, places in the West and South had captured public attention for good, sunlit care. According to Mitman, as bacteriology began to elevate expert, clinical care and regimented treatments, this migration to the sun abated. In short, bacteriology gave rise to a new form of tuberculosis treatment that was largely divorced from nature. This picture is not entirely accurate, for while it is the case that large-scale sanitaria grew popular and doctors proclaimed an unequaled role in treatment, these changes did not mean that nature had no place.[16]

In the 1920s, photographs of children in loincloths on skis constituted a genre of sorts. One image from "the Sun cure sanatarium" in East Aurora, New York, distributed by the Science Service Society, offered the caption "Gungha Din's uniform (plus galoshes and a hair-ribbon) worn amid Alpine snows help these kids to regain health." Other pictures showed children, well and brown, enjoying time outdoors in the sun. Their activities were medical treatments prescribed by medical authorities. Finally, the type of thinking that placed European children at fresh air schools to bask in the sun had become popular in America, and no place defined sunlit wellness more than the sanitarium. The Science Service pressed the point in a picture of three children in loincloths on a sled; its caption read, "These children are not members of a junior nudist camp, but patients in a sanitarium in the high Alps of Switzerland, where they get a maximum of [s]un under near-arctic conditions."[17]

In 1926, *AJPH* dubbed the old sunlight-therapy establishment at Perrysburg "a Mecca for heliotherapy" for its ability to attract enthusiastic American and Canadian practitioners seeking expert guidance. The Science Service provided a photo of a typical sunbath there: children lying on their beds in the snow, with little blankets beneath rather than on top of them. Perrysburg's rigor did not just gain attention from public health officials; lay audiences took note too. *Ladies' Home Journal* pointed to its role in stimulating heliotherapeutic excitement, and the *Chicago Daily Tribune*'s health expert celebrated its devotion to sound practice:

Figure 3.1. East Aurora sanitarium, ca. 1920-40. Science Service Collection, Division of Medicine and Science, National Museum of American History, Smithsonian Institution.

They have kept a very close and scientific record of the amount and intensity of sunlight, particularly winter sunlight. . . . They have made comparative studies of the quality of sunlight in Colorado and some points in other western states with that at Perrysburg. They have compared sunlight passed through dry air with that through moist air; sunlight at high altitudes with that at low.[18]

The vanguard in the nineteen teens, by the 1920s, Perrysburg was no longer exceptional. The United States Department of Labor Children's Bureau took photographs of medical institutions around the country. In one image, a boy lies sunning himself, a cast around both legs and both arms and his head covered. Another, shot in New Mexico at the Tingley Hospital, shows children who, the caption explains, are rolled outside daily to enjoy "a great aid to recovery from many crippling diseases, including infantile paralysis [polio]." Near Atlantic City, at the Betty Bacharach Home for Crippled Children, the little ones are captured enjoying the benefits of a new WPA-built sunporch. Similar scenes were not unusual in the Northeast. In two articles about different Connecticut sanitaria, the *New York Times* described sick children thriving in rugged out-door conditions: in the first, naked kids recuperate by "romping on the beach or swimming in the ocean," and weather is no impediment to their treatment "even when the ground is covered with snow and the ponds with ice." In the

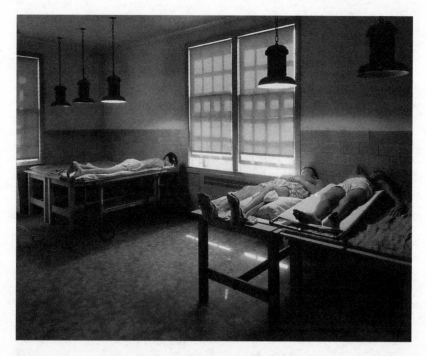

Figure 3.2. Toledo School for Crippled Children, ca. 1920–40. Photo no. 102-G-116-2, Records of the Children's Bureau, United States Department of Labor, National Archives.

second, young Boy Scouts "dressed only in trunks and shoes and carrying haver-sacks, . . . laugh at a thermometer registering 10 below zero and start out in the snow for a real scout hike." The paper reported that the fresh air and sunshine cured nine in ten.[19]

Throughout the nation, for all manner of ills, the prescription was sunlight. In Toledo, it came artificially, with goggle-clad youngsters taking lamp treat-ments. The Marine Hospitals worked hard to provide care for adults too. In one Public Health Service photograph, the New Orleans Marine Hospital shows off its four-person carbon-arc treatment room. The caption makes clear that this was pseudonatural therapy, reading only "Artificial Sunlight." The Veterans Ad-ministration's *Medical Bulletin* discussed its heliotherapy facilities at Fort Ba-yard, New Mexico; Whipple, Arizona; and Rutland Heights, Massachusetts, where the head aide for physical therapy, Mabel C. Ryan, wrote glowingly about the new weapon in the "armamentarium": "The ancient proverb: 'Sunlight is the life of man' is only now beginning to be realized in its whole significance."[20]

In 1932, the American Medical Association set out to study sanitaria. The re-

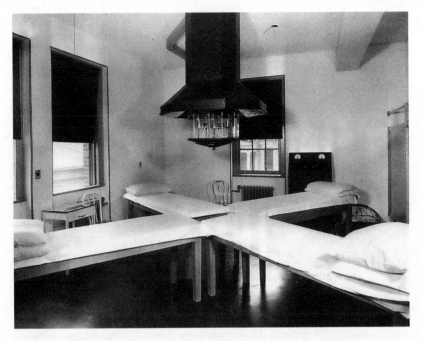

Figure 3.3. "Artificial Sunlight," ca. 1920–40. New Orleans light therapy room. Photo no.
90-G-106-1605, Records of the Public Health Service, National Archives.

sulting report detailed a vast response to tuberculosis, the "great white plague,"
nationwide in scope but varying by facility. There were a total of 95,198 beds
for consumptives, most of them in sanitaria, though some were in wards within
larger institutions. Over a twelve-month period, they admitted 168,818 patients. Since the AMA sought a sense of how forward-thinking American care
was, it counted the number of institutions providing a handful of particularly
promising treatments. Heliotherapy was among the few in this category of "special medical facilities." The AMA findings were positive for natural sun treatments: "Facilities for heliotherapy are quite uniformly presented in the sanitarium group. Most of the institutions have made convenient arrangements for
sun therapy on porches, on roof solariums or in sunpens." When it came to artificial heliotherapy, there was far less uniformity. The majority of institutions did
make some provision for the treatment, generally special wards in the largest facilities and portable sunlamps for the rest. Of the 602 places surveyed, 440 provided artificial heliotherapy—1,589,720 treatments—1,114,709 for suffering
adults and 320,704 for ailing children (the remainder were unclassified).[21]

Though the AMA study found about ten treatments per sanitarium patient, that figure is an average. It is far harder to determine how many tubercular individuals actually received sunlamp irradiation. The famous Trudeau Sanatorium at Saranac Lake in New York State, long renowned for its work treating patients with the fresh air cure, is one of the rare facilities to have offered this sort of statistical accounting. By the 1920s, its director had dedicated the Trudeau program to curing illness with ultraviolet lamps. Between 1922 and 1931, the sanitarium published summary information on the treatment of patients in its annual medical report. Over the period, one in four received a programmatic ultraviolet cure, and only in the first year did fewer than one in five.[22]

Sunlight therapy for tubercular patients had become the pinnacle of responsible care, and state governments proved willing to help deliver it. For example, early in the twentieth century, a handful of men took a trip to South Mountain in Pennsylvania. J. T. Rothrock, commissioner of forestry and leader of the group, noted that the mountain air had done wonderful things for one of his ailing companions and decided that the long-suffering asthmatic should remain in a tent amid the bracing mountain air. In little time, the small site began to evolve, growing bigger with increasingly large state subsidies: after committing an initial investment of $8,000 for a small-scale camp in 1903, the Pennsylvania legislature had agreed to provide $600,000 by 1907. In its early years, the facility was modest: some three- and four-room cottages, a bunch of tents, and a handful of more permanent structures to support care. Before long, it grew to include seventy eight-person Dixon Cottages, named for their designer, the state's commissioner of health. Every inch of the semipermanent structures received sunlight for at least some of the day. By 1920, South Mountain Sanatorium had its own dispensaries, gender-segregated facilities, and a children's wing; within a couple of years, the youngsters received their own preventorium and infirmary.[23]

In 1921, South Mountain was one of three state facilities in Pennsylvania. It had capacity for 1,150 adult patients and a children's wing for 100. The state's two other facilities, Cresson and Hamburg, cared for 650 and 480 respectively. In subsequent years, those totals would change little. Each month, the patients of the Pennsylvania Tuberculosis Sanitaria published a journal of considerable length, which offered articles about treatment, travel essays, works of prose and verse, and updates on facility events. Entitled *Spunk*, no doubt because of the profound importance of a positive attitude and a lively yet relaxing outdoor life, the serial gave a good sense of sunlight therapy's evolution. Even in the small,

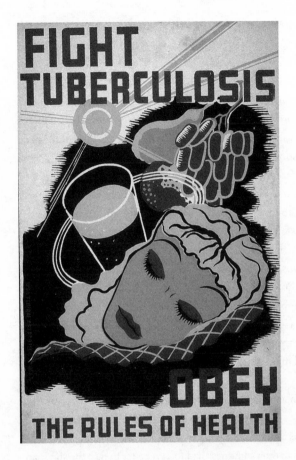

Figure 3.4. "Fight Tuberculosis," ca. 1936-41. By the People, for the People: Posters from the WPA, Work Projects Administration Poster Collection, Prints and Photographs Collection, Library of Congress. Courtesy of the Library of Congress.

nontechnical publication, doctors communicated diverse senses of what good treatment looked like. But there was consensus that science and nature were coming together in an assault on microbes and for the fortification of patients. Over time, there also grew a stronger sense that sunlight was powerful medicine best left to trained professionals.

Though South Mountain boasted that its Dixon cabins were built for light, a day outside did not necessarily mean a day in the sun in 1920. A *Spunk* article by Charles F. Arroyo, physician at Essex Mountain Sanatorium, New York, cautioned patients to stay in the shade especially in summertime. Other articles in the early 1920s encouraged rest and relaxation or moderate activity outdoors but avoided the subject of light entirely. According to one editorial, representative of this thinking, "It is very certain that all the physical universe takes the side of health and activity, wooing us forth into nature, imploring us hourly, and in unsuspected ways, to receive her blessed breath into body and soul, and share in her eternal youth."[24]

By 1922, the same year that editorial was published, the tide had begun to turn slowly toward a more explicit embrace of sunlight. A question to *Spunk's* regular "Question Box" asked about activity while enjoying a healthful sunbath, and the publication responded that, with supervision, it could be okay. Also in that year, Allen K. Krause sang the praises of rest above all, although he did sound a hopeful tone when it came to the prospects of sunlight therapy, pointing to successes using it to treat extrapulmonary forms of tuberculosis.[25]

In 1923, celebrations of the sun in *Spunk* grew more enthusiastic. A rare article or two still featured the outdoor rest cure and avoided any mention of light, but others had a new focus, describing the joy and peace sunshine brought, using the thinly veiled metaphor to dramatize new hope for the dark days of tubercular patients. In September, metaphor gave way to complete and unmitigated mythology in "The Sun Cure," reprinted from *Science Monthly*. Mixing speculative archaeology with enthusiastic zoology, it reached back to the Egyptians and Romans for lessons and made the sun the source of all that was good:

> Old Tut-Ankh-Ahmen, who figures so prominently in our daily press, was brought up as a Unitarian sunworshipper, but later relapsed into the priestly polytheism, which was a pity, for if a people must pick its god from natural objects, as the Egyptians in their blindness had to, it is better to take the sun than to adore cats, crocodiles, hippopotamuses and beetles. The sun is quite literally the source of our vital and mechanical energy, the sole support of all life and motion on the earth, as the ancient Egyptian hymn declares, and we are beginning to recognize, perhaps I should say re-recognize, that it may cure diseases too.

According to this article, Rollier was right, "No tan, no cure," as sunlight stimulated the production of red and white blood cells, coaxing the body's immune system into action.[26]

In a 1924 article, R. H. McCutheon, medical director at South Mountain, offered his own expertise on the matter. He was careful not to overstate the treatment's merits and cautioned that a long initial exposure of the whole body was dangerous—no better than "looking into the muzzle of a loaded gun"—but this powerful weapon could be used for good too. McCutheon lamented that the best cure a sanitarium could offer was to raise patients' immune responses. While some of his theory was still primitive—most notably his celebration of the incandescent bulb as a form of heliotherapy—much of what he said drew straight from the medical literature, including a sense that sunshine grew the body strong, that city life was dangerous, and that there was something to celebrate about the country:

> The effect of light on animal life and especially on man is generally beneficial. We have only to compare the pale anemic child of the city tenement with the red-cheeked, rosy-lipped country child to appreciate what a blessing the one missed and the other enjoyed. Light seems to have a greater tonic effect upon the whole body, stimulating the tissue cells to greater activity and thereby improving the functions of all the organs of the body.[27]

Indeed, 1924 was a good year for heliotherapy in the Pennsylvania system. Articles told of hopeful patients grown strong in sunlight, a newly opened artificial sunroom at the Hamburg facility, and the uniquely and wonderfully sunny climate at Hamburg and South Mountain. The treatment, authors agreed, held a promise far beyond what was occurring in a few sanitaria. It was ubiquitous in nature, with mama spiders putting their babies on their backs and carrying them into the sunlight. In an article reprinted from *The Sea View Sun*, Dr. B. S. Herben wrote that heliotherapy was a health aid for polio and rickets but that it was one that required care. He continued to caution about photosensitivity, as some made poor use of the sun: with heliotherapy, he said, "all we mean is that we are getting Old Mother Sun to do our mending for us. She does it well too, if we give her a chance—that is, we who [can] stand her powerful needles."[28]

Spunk issues over the next few years offered a populist perspective to the enthusiasm for tanning in pieces like the poem "To thee O Sun-God," the travelogue "Leysin," and the popular explanation of treatment "Heliotherapy in Tuberculosis." Authors wrote about sunlight as a disinfectant and as a source of wellness, liveliness, and physical vigor. In the unattributed article "Hail, Helios," the earlier story of Egyptians, Romans, and spiders fused with a celebration of "materialistic science," the "destroyers of beautiful illusion," who had sorted out the true benefits of sunshine. But the real heroes were patients, who, smart and intuitive, brought themselves health. Modern medicine was not the primary focus for "the lowly lounger," who cared little about the science of sunshine or poetry or religion. "[What] the victim of tuberculosis does know is that in the light of the sun lies health":

And so now, as the Spring draws to a close and summer is nigh, daily the worshippers of Helios may be seen prostrate at their adoration. Pale features become first pink, and then tan, and here and there another type of sun spot appears, the well-known freckle. The pestiferous bacillus becomes discouraged
for he can stand far less sun than his victims. He departs, and leaves the healthy,
happy, tanned loafer who soon may give up his life of ease and return to the
struggle of normal life.[29]

Though articles that celebrated smart, sun-worshipping patients and ancients were common late in the 1920s, a countertrend was rising, essays of a
far more technical sort, which took cues from "materialistic science." In *Spunk's*
April 1929 issue, Katharine Blake, executive secretary for the Association of
Catholic Day Nurseries, acknowledged that sunlight, natural and artificial, had
become "one of the most popular topics of conversation" of the day, and that
it could "call forth detailed accounts of experiences and hearsay that [could]
carry one through an otherwise dull hour." Her article, however, was not "Hail
Helios," and she discussed how sun therapy had evolved past vague pronouncements of urban ills to clear critiques of pollution and ultraviolet light–occluding
glass. Moreover, no longer did sunlight simply improve red and white blood
cell counts; Blake's summary was more specific: "Radiant energy is absorbed
by [living cells] and carried in the blood stream, to all parts of the body. When
liberated, it stimulates the intra-cellular processes." Its effects there were mysterious but biologically sophisticated: "Light absorbed by the blood is liberated
as energy in the tissues, but it also has a beneficial effect on the blood itself. It
increases the hemoglobin, which is the coloring matter of the blood corpuscles
containing iron, and also the calcium, iodine and phosphorous content."[30]

Sunlight became somewhat less of a theme in *Spunk* in the early 1930s, and
the articles that did treat it were of a more precise and technically rigorous nature—about specific treatments for tuberculosis of the larynx or bones and
joints (the latter article, tellingly, also featured rest, not sunlight, as the primary
part of its program for treatment). Moreover, vague cautions about the importance of expertise in practice changed to stark warnings. In "Unsupervised Sun
Baths," *Spunk* chided the ignorant lot who spouted off, resulting in too many patients "resorting to the dangerous practice of unsupervised sun bathing." Dangers
could not be overstated: "Such a practice cannot be too strongly condemned,
for any of its harmful consequences in some cases may easily prove fatal." The
caution was clear: "Do not under any circumstances attempt to prescribe for

yourself and to assume the responsibility for results."[31] *Spunk* reflected broader trends, with sunlight medicine transformed over the course of fifteen years from bit player to popular health phenomenon and finally technical subject.

In the late 1920s, Harry A. Wilmer and Lois Parker made a fifty-minute film at South Mountain Restoration Center, revealing a facility committed to heliotherapy that joined medical principle and campy outdoor living. In the film, new patients arrive to a large campus with lots of open-air parlors. Some already-settled residents sit outside wearing only underwear, sunning themselves while another group of about ten hold mirrors and tongue depressors, focusing light on their throats (more on this common treatment for laryngeal tuberculosis later in this chapter). Children play, passing time in games of tetherball, boxing matches, arts and crafts lessons, and on hikes. Other patients seem far more ill: about six girls sit in bed on a sunporch, too skinny and feeble for activity.[32]

This care is not haphazard. In one scene, a doctor enters a children's barracks to examine a group of happy patients. He has them hold out their hands and he looks at their forearms, front and back, to determine the effectiveness of treatment. In another scene, a large number of children lie outside in their underwear with their heads covered. When an employee gives a signal, the entire group flips in synchronicity. Doctors show off their patients' varying levels of physical health and tanness. The facility appears at once restful, physical, playful, and medical.[33]

The heightened public and medical interest in sunshine brought greater enthusiasm for this sort of work, which took children from their toxic homes and exposed them to a vitalizing environment. Health camps were not primarily places for recreation, though according to facility coordinators, children did have fun. They were ways to bring the slightly feeble to renewed vitality. Toledo and Cambridge established camps that found their way into the national press. Ohio's Harrison Nutrition Camp started work with undernourished children at risk for tuberculosis and referred by local nurses. H. E. Kleinschmidt's report on the program's successes encouraged continued funding. Meanwhile, Cambridge could not grow its camp fast enough. Attendance increased eightfold to eight hundred patients between 1917, when it first opened, and 1926, when camp officials began offering heliotherapy to a new type of attendee: the undernourished youngster with a tendency toward tuberculosis because of familial and environmental vulnerabilities. This was the new preventorium model, with roots in older American institutions but more like Rollier's vision. South Mountain had one; its goal was both educational, to teach good health habits,

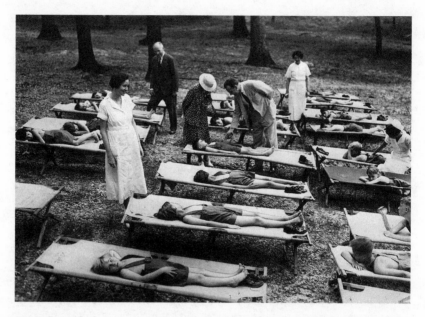

Figure 3.5. Health camp, n.p., n.d. Photo no. 90-G-121-2913, Records of the Public Health Service, National Archives.

and constitutional: "A maximum of fresh air and sunshine gives an increased resistance while graduated exercise and supervised play insure firm muscles and not fat."[34]

The new tendency was to see diseases holistically, as the result of conditions that wracked the weak but did not affect the strong. Indeed, bacteria could do damage, as could a nutritional deficiency, but health regimens could prevent both. So why stop with the hospital?

In an advertisement on July 27, 1927, Vita Glass boasted that New York's Equitable Trust Company building had recognized the value of ultraviolet rays and had glazed for sunlight. The entire right-hand column of the promotion celebrated institutions that employed the product; most were hospitals—including the Desert Sanitarium in Tucson, the Pokagama Sanatorium in Minnesota, and Good Samaritan in Cincinnati—but in addition to the Equitable, buildings for L. Bamberger, Eli Lily, and E. I. Du Pont de Nemours stood out. The accompanying text made the case for other businesses following suit: "In this highly competitive era, the efficiency of a business organization is in ratio to its health." Other promotions boasted that executives had installed Vita Glass because "the sedentary life of the modern businessman makes it more necessary than ever before to take every step to guard his health." Still others provided thumbnail-sized drawings of places the new product had a use — often the factory was one of them. Vita Glass was selling a healthy workforce.[35]

In 1930, with the sunlight movement in full swing, newspapers across the nation commented on the erection of what they claimed was the first window-less factory, in Fitchburg, Massachusetts. Built by Simonds Saw and Steel, the facility was only practical, according to realtors, because it could provide ultra-violet light artificially. One member of the New York Architectural League carried the argument further. Windowless plans were preferable, he said, because of better temperature control, quieter conditions, more uniform light, and "the increased amount of ultra-violet rays possible through the use of lamps rather than through intermittent sunlight."[36]

Like other forms of welfare capitalism, though, the goal was as much heightened profits as happy workers. Sanford DeHart, director of the Hospital and Employment Departments at Cincinnati's R. K. Le Blond Machine Tool Company, made the case for sunlamps most aggressively. He cited a Boston study that found that the average male employee lost 1.4 days per year to colds and the average female 2.1 days. The cause of these illnesses was, in part, the confinement of industrial workers when they might otherwise have been outside enjoying the sun's rays. That led DeHart's Company to install Cooper Hewitt Lamps on the shop floor to replicate June sunlight. The health expert told his readers that management was not alone in advocating such measures. In an effort to improve worker health, the Garment Trades Workers had established a cooperative in New York that provided treatments for a nominal annual fee of one dollar.[37]

These references to particular businesses making use of light therapy were far from exceptional. An article in *Architectural Forum* noted Philadelphia's Watson Stabilator Company, and the *New York Times* and the *Washington Post* reported that the United States House of Representatives had found a new use for tax-payer dollars: a sunlamp in its shower room. But it was in places like mines, where there never was and never would be sunshine, that capitalists looking for a productive workforce could employ artificial nature to its fullest benefit. One article in *Engineering and Mining Journal* agreed that the darkness problem in industry was greatest for those who worked underground, "where conditions of employment unavoidably deprive the workers of practically all of the health value of natural sunlight." Trapped beneath the earth, the article averred, employees succumbed to respiratory infections at an alarming rate. Though it seems likely that miners fell ill for reasons other than inadequate light, here the solution was a special four-bay light-treatment room. At least one firm took up the call. Idaho's Bunker Hill and Sullivan Company chose a different installation, but it did offer ultraviolet radiation to its workers beginning around 1923. The enthusiastic response led to a larger installation in the late '20s, which would also treat miners' families if they wanted the benefits of sunlight. A report in *Mining and Metallurgy* claimed that a 50 percent reduction in colds and respiratory ailments, similar to the results achieved at Cornell, seemed a reasonable goal.[38]

Hard work needed hardy souls in good health, but that was not the only physical requirement for jobs done well. Matthew Luckiesh articulated the sunlight needs of workers frequently. The long-time General Electric employee, who had risen to the head of the Nela Park National Lamp Works, showed a profound concern for ultraviolet light but added another layer to the needs of workers: the specific quality of daylight for visual tasks.

Luckiesh's rationale for his theories was biological, his arguments scientific. He said he was uninterested in debating medicine—that was not what he knew and, besides, the discussion was unnecessary. Humanity was natural, an inexorable truth and unavoidable reality. Evolution had written certain needs into bones, bodies, and eyes. Coming indoors could not overcome that in generations; life inside was simply contrary to human nature. In order to better craft their environment, people had to recognize those facts and adapt the indoors to simulate the out-of-doors. In his book *Artificial Sunlight*, Luckiesh argued that it was egotism that led man to attempt to sort out what about nature was good or bad. Instead, he should take lessons from "the blade of grass," which "devoid of a means of locomotion is protected from making mistakes. It remains in the environment where it was born and to which it is adapted. Who can say how much human beings have blundered by leaving their environment?" Fortunately, Luckiesh, part scientist, part naturalist, wholly optimist, saw a solution. Though civilization had severed people from their natural, healthful environment, recently, it had also offered a new hope.[39]

For centuries, content with the world it created, humanity had done little to transform it, but that was no longer satisfactory. For Luckiesh, civilized life was indoors, a problem as long as primitive illuminating technology made the best and most reliable source of light daylight. In congested cities, the sun was a cruel master, needing huge windows, and handicapping architecture with walls of glass and setback requirements. But by Luckiesh's time, the future was bright, because of the resourcefulness of man, "the only animal that is continually evolving in independence, the only one who is inoculated with appreciation and defiance of Nature, the only one who worships and challenges the Sun." With the realization that "nature could be understood and was worth understanding" came an awareness that modern science could free humanity from the chains of its natural environment. With careful study, innovators like Luckiesh were unraveling the mysteries of the outdoor world and discovering sunlight's spectral characteristics. For the far from impartial author, "no art excel[led] lighting as a benefactor of civilization."[40]

Luckiesh's theory brought him to a new "science of seeing." Though his conclusions passed from empirical to speculative, he clearly thought that the basis of his thinking was rational and biological. Ultraviolet energy—or radiant energy, as he called it—was important, but human sight had other requirements: like the rest of the body, the eye belonged outdoors and had needs best met in nature. Having evolved to look at things in the distance parallel to the ground, people suffered when they tried to complete tasks like reading to which they were unsuited: evolution had produced hunters not scholars. The discomfort attending these unnatural visual acts was exacerbated by bad lighting. Good lighting and efficient seeing required attention to seven characteristics, including brightness, contrast, glare, softness, and reflected glare. Citing a unit of measurement widely used in the lighting industry, Luckiesh argued that outdoors the sun provided as many as 10,000 foot-candles; indoors, usually less than 20. Outdoors, a single overhead source gave ample brightness; indoors, that was not the case. Outdoors, atmosphere diffused light making it uniform; indoors, point sources caused bad shadows. Daylight through windows was not much better than crude bulbs. It was not natural, streaming in on one side of the room, casting shadows and yielding bright spots and dark patches. Humans were "seeing machines," executing age-old processes that required the careful coordination of "nerves, muscles, organs, bodies and minds." Without proper lighting, strain reverberated through the human mechanism, yielding exhaustion, muscular and nervous tension, and even destabilizing the heart rate.[41]

Luckiesh expected to accomplish his goal of good seeing underneath good light. He titled one article "The Color of Daylight"; another celebrated fluorescents for, among other things, their similarity to nature. Researchers, he thought, would precisely measure the amount of light needed to accomplish specific tasks. Technological innovations like photometric devices, combined with the expertise of physiologists, psychologists, and applied physicists, would complete the picture, yielding a future better adapted to human needs.[42]

The 1933–34 Chicago World's Fair was a place for the zealous enthusiasm of scientists like Luckiesh. Though it came at a tough time, with the nation mired in the Depression, the Century of Progress exhibition—commemorating Chicago's hundredth anniversary and linking the city's rise to wondrous growth in innovation—spoke about hope. The windowless Hall of Science fittingly testified to the promise of a more perfectly engineered future. Lighting experts, the official guidebook noted, offered a "practical new design" to avoid "the variability of daylight," with "constant control over the volume and intensity of light." The steel-and-glass House of Tomorrow—in some ways the science hall's opposite—also celebrated the new bright opportunities in innovative plans and building materials. It too indicated the possibility of a more fully managed environment because, though bathed in natural light, the modern home had not a window that opened. A huge hall with no windows and a dwelling that was all windows sealed tight suggested the abounding prospects of a better-planned world. In the *Chicago Daily Tribune*, Philip Kinsley's article "Human Miracle of Light Enters New Era at Fair" told of the hard-working engineers who put together displays that marked "the beginning of a new period of outstanding achievement in light engineering." Luckiesh, no doubt, was proud.[43]

Schools, with their necessary reliance on visual tasks and concern for the welfare of America's youth, operated at the intersection of workspaces, homes, and hospitals. Luckiesh concerned himself with the peculiar nature of daylight coming from the sun; it formed the backbone of his theories. At times, he took his thinking into the classroom, and a few others joined him. In one article, W. R. Flounders argued that one of the reasons children struggled to see at school was because lightbulbs did not match sunshine through windows. The difference strained sight. But for the most part, studies of lighting in classrooms criticized natural light as too variable, contingent on fickle sunshine that might not make it to the earth.[44]

Unfortunately, it is often hard to determine whether lighting experts were primarily worried about the quality or the quantity of their product—whether the main issue was the specific nature of sunshine or simply levels of brightness. In a picture that the Public Health Service took at the Antietam School in Hagerstown, Maryland, the concern was clearly brightness. A wall of windows, all open, glows bright in the sunlight. The caption on the back explains that the school represents the most modern of well-lit buildings, but many illuminating experts would not have agreed. Luckiesh worried about the uneven dispersal of brightness throughout a room and noted that windows distorted classrooms, making them long and thin. He believed that if light could not penetrate deep into a building, it could not reach all students in sufficient quantities, and that was a profound problem. The Antietam School design did not necessarily respond to sunshine concerns, and neither did the many other plans that flooded rooms with artificially produced brightness with little regard for its proximity to nature. These designs did not accord with the science of seeing or address Flounders's concerns about the difference between light generated by indoor bulbs and light generated by the sun. Illumination experts had moved past the early fresh-air-school age when all sunshine was equivalent, when light through windows was sufficient, and when brightness was key.[45]

In the treatment of "feeble" students, sensitivity to sunlight specifics now prevailed. Rollier's model École au Soleil took children at risk for illness into the country and educated them where, he believed, they would develop best. According to Vita Glass, one London school offered the test case for a similar program, albeit without the country air and mountain climate. Children at the Smethwick School, aged nine to eleven, took their studies in Vita Glass classrooms. According to the company, boys gained an average of 2.83 pounds and 1.22 inches more than their compatriots in regularly glazed rooms. They also made more red blood cells. Girls faired even better, 6.11 pounds and 1.86 inches. With such a stunning resume, it was little wonder that Vita Glass scored successes in selling to other schools. According to an article in *AJPH*, when Bronxville decided it would assess the merits of specially glazed classrooms, Vita Glass stepped up to offer its support. In other publications, the company advertised its good work, boasting installations in Cleveland; New Britain, Connecticut; New York City; Chicago; Pottstown, Pennsylvania; and Cambridge, Massachusetts. Indeed, the preoccupation with sunlight ran deep in Cambridge's school system, the primary feeder for the city's health camps.[46]

Lustra-glass's product specification sheet featured schoolrooms even more proudly. It claimed to be a pioneer in this kind of work and pointed to buildings glazed with its product in Clinton, Iowa; Terre Haute, Indiana; Buffalo; and Milwaukee. Here, the color of daylight did matter; Lustra-glass, as its name suggested, was concerned with the clearness of its product. The company glossed over its prior confession that its windows would not transmit the lower wavelengths of ultraviolet light: Northville Grade School in Michigan was described as a "Truly modern school" employing "a truly modern windowglass," so clear that it "transmits more daylight and more of the healthful ultra violet rays of sunlight."[47]

Schools that installed such glass were enthusiastic about the benefits it bestowed. Los Angeles's Urban School featured its use of "ultra-violet health glass" in advertisements. The boy's military academy, which boasted a rich, outdoor experience combined with classwork, claimed that it was the only school in America that used the special windows in all of its rooms. The local press picked up the school's boasts, claiming that this was a remarkable place, properly sunlit throughout, which brought to its young students unparalleled benefits. In one of its articles, the *Los Angeles Times* celebrated the "special progressive feature in which Urban leads" as one of the reasons that children there had such a strong health record.[48]

At the Third Race Betterment Conference, John W. M. Bunker reminded his audience that ancient civilizations had celebrated the sun without knowing why.[49] Yes, the Greeks had solaria, he conceded, but "today we have open air schools and sun rooms glazed with a special glass that permits the passage of ultra-violet light." Inclement weather represented no obstacle in his argument: "We press a button and at once have at our command an electrical source of these vital rays stronger than that of the noon day sun in an unclouded summer sky." As the Urban School's boasts indicated, Bunker was overselling modern successes; most school districts could not afford artificial sunlight for all students, all of the time. More common were individual facilities like Newark's Boylan School, or rooms within schoolhouses, like the one in the Osborne Street Grade School in Sandusky, Ohio, where "crippled or otherwise physically defective children" received hour-long treatments beneath nine S-1 sunlight units. Berkeley's Sunshine School, a descendent of the fresh air school, offered outdoor wellness all year. Each morning, undernourished but otherwise healthy youngsters donned their sunshine suits and endured a rigorous schedule of an hour of class, an hour of sunbathing, and an hour asleep; the afternoon program was similar. And the facility, according to enthusiasts, was a success: children experienced greater pep, improved appetite, academic progress, a more developed musculature, and decreased nervousness.[50]

Two of the most arresting examples of the sunlight movement came from schools; together, they dramatized the curious new naturalism that was emerging. In 1935, an article by Sally Joy Brown celebrated Chicago's Spalding School for Crippled Children, the best such institution in the country. It was a model and marvel to which people in every state, as well as London, Berlin, Montreal, Vienna, Buenos Aires, Tokyo, Shanghai, Sydney, and the British Indies traveled. Built in 1928, it had since acquired a first-class solarium, featured in a 1929 issue of *JAMA*. The solarium did not open onto sun or sky; it was not a sunroom but a large machine. Children in loincloths—made in home economics—entered the device on one side and were moved by means of a conveyor belt past six artificial sunlamps. The belt, which doctors could slow or speed to better tailor treatment, standardized light dosages. The children were products

Figure 3.6. Light therapy at a school in Westchester County, New York, 1940. Science Service Collection, Division of Medicine and Science, National Museum of American History, Smithsonian Institution.

on an assembly line. This solarium used a perfectly regulated but most peculiar sunlight.[51]

By 1940, sunbaths at a Westchester County school had eliminated sunlight entirely. In one photograph from the Science Service Collection, it is hard to read where exactly the children are; they are not dressed in loincloths or wearing goggles. Instead they sit at a table, in what appears to be a room lit only by a sunlamp, building strong bones under the artificial light as they drink special vitamin D milk.[52] The caption the Science Service supplied for a related photograph of children playing with blocks emphasizes that at this innovative institution sunbaths have done what winter sunlight cannot and that these children have managed to secure strong bones in minutes with a stronger-than-the-sun lamp. This was far from Berkeley's Sunshine School. As was the case at the Spalding School, nature was entirely absent and the sun was no longer necessary.[53]

Figure 3.7. Light therapy at a school in Westchester County, New York, 1941. Caption: "Building with blocks and vitamin D goes on simultaneously in this Westchester nursery school where sunlamp bathing is a regular habit. While the children are building with blocks, their General Electric sunlamp, with its high output of anitrachitic ultra-violet rays, is helping the body build vitamin D and stronger bones and teeth." Science Service Collection, Division of Medicine and Science, National Museum of American History, Smithsonian Institution.

Sunshine enthusiasts saw no need to limit their efforts to the wellbeing of young children or white-collar workers. Though it might not have had the great effects advocates thought, natural light in the treatment of zoo animals made sense given the history of experimentation. Many of the early studies, from Mellanby's mistaken work on dogs to tests on the rachitic bones of rats, used animals as subjects. Thus, it must have been unsurprising that Vita Glass felt confident it held the answer to the London Zoo's health problem. With caged animals looking sluggish and taking ill, the company began glazing cages for sunlight. It reported substantial successes. In 1927, a Vita Glass ad in the *New York Times* informed readers that veterinarians had swapped out a large pane of glass in the lion cage. The animals had of course gravitated toward the full-spectrum sunlight, with the results of better spirits, fuller coats, and more robust appetites. That first success led zookeepers to expand the program, and soon the lizards, too, had started feeling better.[54]

Such reports also helped spread the use of lamps for animals. In the discussion of his paper delivered in 1927, J. S. Hughes of the Kansas State Agricultural College, wrote hopefully about the future of zoos,

> Since the difference between glass filtered and direct sunshine has been definitely understood, it has been possible to keep many animals in captivity in a thrifty condition which could not be kept in confinement before. In some cases in the zoo the "vita" rays are obtained by a special type of lamp while in others glass substitutes which will transmit the "vita" radiation of the sunshine are used.

Hughes was not at the cutting edge of zoology with his pronouncements. In that year, the Bronx Zoo began lamp therapy on a few patients with "cage paralysis." By 1932, the work had expanded to help Buddy, a begoggled chimpanzee; the National Zoo's baby gorilla N'Gi was feeling better from his cold; and veterinarians in Pittsburgh and San Francisco were treating their caged animals similarly.[55]

It was on the farm, however, that animal treatments with light therapy seemed most promising. In 1932, Irving J. Kauder, from Edgewater Farms, New Paltz, New York, wrote to Semon Bache and Company to tell of his successes with its ultraviolet glass. The 3,102 Leghorn chicks on his farm had faired exceptionally well: lowered mortality, stronger legs and bones, more luxurious feathers, bigger bodies, and fewer diseases. Kauder expected to install the window substitute throughout his brooders. The experiment, a joint project between farmer and glass pioneer, also produced far more eggs than usual. In marketing literature, Semon Bache implored, "Let Nature help!" Its SUNLIT glass was a cost saver: "Nature's free sunshine" allowed farmers to avoid expenses like artificial light and dietary supplements. According to the pamphlet, doing nothing was not an option, as farmers had to seize the opportunity for the related benefits of increased egg production and higher profits.[56]

The rationale for sunshine in poultry farming had been established years earlier. In 1924, the *Chicago Daily Tribune* reported that a Kansas agricultural station had proven that chicks bred in darkness died. A year later, the Hanovia Chemical Manufacturing Company's pamphlet *More Money in Eggs and Chicks* promised healthier chickens laying more eggs with stronger shells. Hanovia could have been writing about humans in extolling the benefits of its product:

"Ultra-violet light is the greatest chemical force of nature. . . . It is the force that controls the production of vitamins in all animal life. Without it vegetation and animal life cannot thrive." General Electric also articulated an expansive vision in which there could be no alternative to innovative agriculture, describing a "new day in scientific farming and in the poultry industry." To ignore change meant ruin; competition would crush those hemmed in by tradition. According to the *Chicago Daily Tribune*, the corporate giant was wrong; for most farmers, artificial sunlight was too expensive. Still, there were some who, the newspapers reported, had tried to secure healthy animals by creating sunshine for them.[57]

In dairying, enthusiasm was similar—if a bit more measured—for healthier cows and for milk rich in vitamin D. Interested in securing a bigger market for his product, a representative of the National Carbon Company told an assembled group of illuminating engineers that radiation was useful and effective for the dairy industry, but only if practitioners focused it on bovine bellies, for their backs were simply too tough.[58]

Within a decade, sunlight concerns had mushroomed across the board. It is important not to minimize the depth of earlier housing reformers' worries about urban conditions, but their light concerns were limited compared to those that followed. Research had revealed a new set of threats. Doctors and scientists asserted their role in checking them, but experts were not the only ones in these debates. That was a problem for recognized and accredited authorities who did not like sharing the stage. They began arguing that dangers came not just from darkness but also from ill-considered treatment. *Ultra-Violet Rays*, Percy Hall's extensive and engaging celebration of sunlight, cautioned that too much of a good thing could be dangerous, although the warning was buried amid chapters explaining the benefits of artificial sunlight for conditions ranging from malignant diseases to nervous and metabolic conditions. This was a typical approach, to warn about overuse while encouraging long-term treatments for myriad conditions. Clearly, the medical establishment struggled with the dissemination of information. While it could brand as irresponsible the claims of overly enthusiastic purveyors, limiting overexcitement and misinformation within its own ranks was not as easy.[59]

The dangers the AMA saw were many and real: overexposing the sensitive tanner could damage the skin, leading to atrophy, splotchy pigmentation, and even cancer. For "normal persons," excessive tanning could kill, and localized burns—though useful in some conditions—could permanently damage the skin. Other sunlight experts wrote that the benefits from therapy reversed with overexposure: bodies became more susceptible to disease and less able to heal themselves. Practitioners rightly recognized the danger of bad burns, but with sentiments like Percy Hall's common, many in the profession were unsure about just how hard doctors should push to limit a treatment that they firmly believed had a place in each human's life.[60]

Sunshine prescriptions were supposed to be based on careful, clinical observation, crafted by experts who would evaluate patients' needs and develop tailored treatments for better health. Authors sought to distinguish accurate medical claims from those of "quacks," "fakers," "nostrum vendors," and others "reaping a rich harvest from a gullible public." Such tricksters "[lay] in wait to attack the public welfare and extract hard-earned money from their dupes, and in every way [used] all the means at their disposal for victimizing the public at large." Critics believed there was plenty of fault to go around when it came to poorly considered treatment. Ill-informed sunbathers thought they knew what they were doing. Incompetent doctors encouraged patients to spend long hours beneath a sun too strong for their unprepared skin. Bad scientists produced dime-store studies that led self-styled experts to prescribe treatments that did little good and could cause much harm. And lamp makers' extravagant claims duped an unsuspecting public looking for a quick fix that was also a certain cure.[61]

Although the AMA did not withhold blame from any group, the key problem in its view comprised poor treatment, excessive enthusiasm, and bad machines that gave false hope. Its 1936 *Handbook on Physical Therapy* discussed light treatments at length, offering five sections on ultraviolet radiation or sunshine. With a general message of caution, the official AMA statement agreed that ultraviolet light was good for most forms of tuberculosis and that it was probably of benefit in treating wounds; moreover, though assertions about its effects in cases of anemia were overstated, it might be some help as an ancillary treatment. However, the AMA strongly challenged claims that ultraviolet light had a benefit as a general tonic.

Both because it was no doubt difficult to take aim at accredited doctors themselves and because misunderstandings often stemmed from irresponsible, profit-hungry lamp makers, the association took primary aim at industry: "Not satisfied with their triumphs thus far, these manufacturers began to advertise their machines directly to the public and many were the unwarranted therapeutic claims made for their use in every known disease."[62] Such chicanery was all supposed to have ended back in 1932, when the AMA decided to oversee new sunlight technology. In order to gain the association's approval— and manufacturers enthusiastically advertised it when they did—lamp makers had to agree to meet a new set of requirements. Advertisers could only claim that "sunlamps" helped bone and tooth development and treated rickets. "Ultraviolet sources," which produced a stronger shortwave spectrum, were medical tools distinct from sunlamps that did not belong in lay practitioners' hands. Advertisements could only profess to treat the conditions that the Council on Physical Therapy approved. All artificial light lamps had to stick to reliable and verifiable claims about the radiations emitted, and offenders would find their products on a list of unacceptable devices. In this process, a form of regulation emerged; bad products were blacklisted while good ones could advertise their medical sanction to the public.[63]

In subsequent issues of its journal, the AMA approved lamps that made modest claims and rejected those that did not meet specifications. By 1933, the rules had tightened and clarified; a sunlamp was defined as a device that

> at a specified distance, emits ultraviolet radiation not differing essentially from that of the clearest weather, midday, midsummer, midlatitude, sea level, natural sunlight, in total intensity and in spectral range of wavelengths extending from about 2,900 to and including 3,130 angstroms [10 angstroms = 1 mμ], and that does not emit an appreciable amount of ultraviolet radiation wavelengths shorter than 2,800 angstroms.

These machines, with their lower risk of overdose, were distinct from therapeutic ultraviolet generators for professional or institutional use. The AMA found, defined, and regulated real sunshine, the type manufacturers should replicate.[64]

General Electric tried to comply with its S-1 and S-2 lamps, which were rewarded with approval. The theory behind the S-1 type was dual-purpose lighting, combining mercury vapor for health and tungsten filaments for sight. GE claimed that the S-1 was an epochal lighting source—to physicists, as "close to natural light as we've come." A 1933 paper celebrating the new dual-purpose-lighting concept admitted that it was not therapeutic; it would supplement, not replace, more powerful ultraviolet emitters. Indeed, the AMA argued, true dual-purpose illumination was impossible in most cases. Lamps that provided ultraviolet light had to shine near to their subjects, which was far from ideal when it came to lighting an entire room: units that only worked at thirty to thirty-six inches from their users and directly overhead—the recommended specifications for most sunlamps—made little sense.[65]

All parts of the debate over sunlamps were fractious. Some authorities were enthusiastic about the prospects of dual-purpose lighting; others thought the goal was foolishness. Some thought sunlight brought general wellbeing; others thought its effects were limited. Some wanted to put lamps in the hands of the public; others thought the dangers of such widespread treatment were too great. The AMA tried to close ranks, but contrary opinions continued. It could try to control manufacturers through a certification process, but it proved harder to silence dissenting doctors. The zealous claims of sun-happy medical practitioners continued well after the AMA issued its statement attempting to curtail the cure. Sometimes historians of medicine try to draw a firm line between patent medicine and professional doctrine. With sunlight, there was no firm boundary, as the profession failed to control a group of experts who claimed their natural practice was firmly rooted in sound science.

With the rules of effective sunlight treatment yet to be defined, an enormous variety of interests entered the debate, and inventors frequently rethought the ideal tools for capturing health. They believed—and doctors agreed—that the sun was natural and good, but these forward thinkers began to wonder if they might be able to do better. They sought an understanding of nature's gift that would enable them to make products that improved sunlight.

Some, certainly, called this vision folly and nature ideal. They said, as did the authors in *Spunk*, that time basking in sunshine was part of a total experience that made a day outdoors great. In the words of one *Hygeia* editorial,

> The Outdoor breeze, the exercise, the fresh air, the stimulation of pleasant companionship on the golf course or in the walk through country lanes are factors which no artificial source of ultraviolet can even suggest and which certainly cannot be compensated for by any other means. . . . Let him get not only a sun tan but the relaxation, the joyousness and the inspiration that come with green fields and the hills and the blue skies and sun above.[66]

But for another, less rigid, class of sunshine enthusiast, it made little sense to wait for or try to chase down the fickle sun. Often this group acknowledged that natural light was preferable, but it was too unreliable; the environment could be a formidable enemy. Sickly children, deeply in need of a regular regimen, could not wait for clear skies, a trip to the mountains, or the summer sun. This was the argument for conservative artificial sunlight advocates: lamps to pick up where nature left off.

In the 1936 *Handbook on Physical Therapy*, Henry Laurens contended something a bit more forceful: the objective for new treatments was a "spectrum of sunlight found most efficient in the particular condition it is desired to alleviate or aid." In other words, he did not want new or unnatural light, but neither did he seek to replicate the full solar spectrum. His dream was a more precise parsing of nature's benefits. Doctors could handpick their sunlight, in essence outsunning the sun by isolating and then replicating healing wavelengths. According to Frank Krusen, "Many clinicians have noted that we must be more discriminating with our use of rays. We should target the spectrum to the patient, much as we target drugs to patients."[67]

In the 1920s, the goal in reproducing ultraviolet light was to recreate the sun's health-giving property. That could be done with great facility. In time, ultraviolet radiation gained its own identity—a type of light found in sunshine but also found elsewhere. There was no transitional moment for that change, and even those clinicians who chose a spectrum markedly different from sunshine's continued to consider—and call—their light natural, sunlike. Healthful sunlight had become synonymous with handpicked ultraviolet light. In making these moves, experts created a third option to the conventional natural/unnatural binary; it was, for lack of a better word, hypernatural. It took its inspiration from nature but moved beyond it, creating a more perfect cure.

It was very hard, for example, to treat tuberculosis of the larynx with direct sunlight—until the laryngoscope came into use. By affixing reflectors onto a prop that attached to the back of a chair or a patient's head, clinicians could focus sunlight into his or her throat. The result was a direct application of rays to the site of infection. This treatment mirrored another. In General Electric's *Manual of Standardized Operative Technic for Users of Victor Ultraviolet Lamps*, the company claimed that its product was better than sunlight because it eliminated heat and was more reliable, working regardless of the weather. With special attachments, like the vaginal and rectal quartz speculums, the rectal and vaginal applicators, the prostatic applicator, and the laryngeal applicator, the Victor Water-Cooled Mercury Lamp could bring ultraviolet to new parts of the body. The technology's considerable benefits brought competition, most prominently from the Cold Quartz Company's Kromayer applicators.[68]

All the best parts of sunlight, according to enthusiastic reports, were going where the real thing never could. With sooty skies, dense atmosphere, and glass windows choking off solar wavelengths, lamps could reasonably claim to replicate a light more natural than nature's. Sunlight was imperfect, and humans, with the unintended ills of their civilization, had made it far less perfect. Lamps brought the prospect of "clearest weather, midday, midsummer, midlatitude, sea level" light to the entire planet, but why stop there? With applicators and devices that emitted narrow bands of the solar spectrum at high power, tech-

MISC. APPLICATORS AND ACCESSORIES

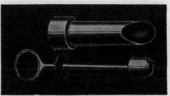

Quartz Speculum-Vaginal. Cat. No. J-6125. Designed by Dr. G. C. Wagner of Tacoma, Wash., for use both as an aid to diagnosis and for the application of ultraviolet light in vaginal cases. Being made of pure rock crystal (quartz) it is transparent to ultraviolet. With the speculum *in situ*, the physician may use any of the proper applicators to direct the rays within the speculum to the particular area requiring exposures. The tissues being under compression are rendered anaemic; which makes for better penetration and more efficient bactericidal action.

Prostatic Applicator. Cat. No. J-6115. A metal obturated speculum of convenient dimension to insure an efficient distribution of ultraviolet energy over the prostatic region when introduced rectally.

Wagner Prism. Cat. No. J-6108. Large (G. U. and rectal).

Quartz Speculum-Rectal. Cat. No. J-6126. Same as the above but intended specially for rectal work. This instrument will be found of marked value in the observation of treatment progress in hemorrhoids while under ultraviolet irradiations. It is often used in the treatment of prostatic cases.

Quartz Laryngeal Applicator. Cat. No. J-6122. 6½″ long with right angle curve and so constructed that the light emission takes place almost entirely at the tip. Useful particularly in the treatment of laryngeal tuberculosis.

Figure 3.8. Page from *A Manual of Standardized Operative Technic for Users of Victor Ultraviolet Lamps*, 1929. Victor X-Ray, General Electric Trade Literature, National Museum of American History, Smithsonian Institution Libraries.

nology could offer something even better than nature's best. Innovation was finding new ways to replicate all of the sun's rays or to produce those special wavelengths that people needed.

In 1927, *Hygeia* published a fictional story, "The Great Sky Medicine," about a pair of Indian children. In winter, the girl, Windflower, comes down with a mysterious constitutional ailment brought by "the hunger wolf." Enfeebled, she needs help. Others near her suffer similarly until the village enlists the help of a crow who flies up and pecks a hole in the sky. The best part of this story, according to *Hygeia's* narrator, "Any little girl who droops and fades, as Windflower did, can be made well and happy by the great sky medicine, the big, warm, wonderful sun."[69]

Nine years later, *Hygeia* printed "Unfortunate Egglebert Ploot," the tale of a sickly, pale, bookish boy. Tired of watching him suffer, Egglebert's friends steal an airplane and take him to the Sun Bakery, owned by Violet Ray. There he is put into ovens and baked until golden brown and made terrifically strong. Fortunately for unfortunate Egglebert, science can do nature one better.[70] "The Great Sky Medicine" offered a campy naturalism far different from the sentiments in the later story. In both, the goal was sunshine, but the means of transmission had changed drastically. Nature was largely written out of the 1936 picture in favor of a more perfect, technologically sophisticated treatment.

Sun therapy had spread in professional hands. It reached hospitals and zoos and, according to its many advocates, had cured the sick and fortified the feeble. Some authorities claimed that heliotherapists and electrical engineers had become irresponsible with their enthusiastic pronouncements, but it was not easy to contain a medical movement legitimized by apparently sound science. Treatments sought the delivery of a hypernatural gift with great care and precision.

For those who lamented that the urban, industrial, indoor lifestyle was dangerous and for doctors who called for more time in the sun, the line between natural and artificial was blurry. Their solution was not simply a day sunbathing in the country. They saw the importance of dosage and careful prescription and sought new innovations to return balance to sun-starved Americans. Their faith lay in science and technology. These treatments did not simply encourage sickly patients to throw off their modern trappings and head back to nature. Nevertheless, that was all that some people wanted to do. They had their own sunlight agendas and offered their own solutions. It was on beaches, in bathing suits, or back at home that everyday folks sought their sun.

4

Popular Enthusiasms

Eugenists, Nudists, Builders, Modern Mothers, and the Sun Cult

In the winter of 1928–29, King George V of England was suffering from a pulmonary infection, and Chicago's health commissioner, A. H. Kegel, claimed to know why. London, he told the *Chicago Daily Tribune*'s readership, was a sooty place; its atmosphere shut out light, and as a result, the monarch was near death. In an effort to stem the illness's tide, the king's doctors had given him an artificial sunlight bath to reproduce the vital health element. In subsequent weeks, *Tribune* readers and those who regularly picked up the *New York Times* would hear a lot about the king's condition in page-one articles updating his progress and telling about the artificial sunlight he enjoyed. With these good treatments and recuperation in country sunrooms, the monarch survived. This story was one of myriad places Americans learned about the health benefits of heliotherapy.[1]

Medical professionals largely drove early sunlight treatment, but before long, word spread about its contributions to good health. Undoubtedly, few people were reading medical treatises, and many avoided a stay in a sanitarium, but there were other ways to find out about the sun cult, and lots of people were interested. In the mid-1920s, information began to appear in newspapers, magazines, and pamphlets from government bureaus, encouraging a form of practice less regimented—but still well organized—than what doctors counseled. The stream of news exhorted mothers to tan their children and men to undo their collars and shed some clothes. It told poor people to head into the sun and encouraged white-collar workers to step away from their desks and go to the beach. Some of this discourse targeted women, who were most likely to dress in a more healthful manner and who, as stewards of the home, had exceptional responsibilities. Other arguments were tied to an emerging trend in eugenics. These contentions either cautioned that creeping darkness threatened the white race or justified a new aesthetic ideal in which brown was beautiful.

Before long, lamps were not the primary sources of sunlit healthfulness and a trip to the beach had become an unmitigated good. Some of America's sunniest spots capitalized, claiming that they alone could unlock the fountain of youth. Then there was nudism, an encompassing, though not always fully reasoned, philosophy, which argued that if fewer clothes were good, none were undoubtedly better. It called upon Americans to show their skin and enjoy time outside. This thinking, some homegrown, much imported from Europe, claimed that nudity meant health, mental and physical, and seized upon the newfound merits of sunbathing as evidence.

Americans did not have to rely on the crass ads of glass pushers to get their news about sunshine—neither did they have to visit the doctor. Experts with regular columns in major newspapers kept the nation well apprised of the sunshine craze. In W. A. Evans's "How to Keep Well," a regular series in the *Washington Post* and the *Chicago Daily Tribune*, readers heard about the theory behind the new sun treatments.

Back in 1915, Evans had begun touting the benefits of sunlight: "If a person is so closely confined during the year that he develops house pallor he should take advantage of his vacation to give his tissues the advantage of sunlight." Evans further pointed out that ultraviolet light was most important to wellbeing, that pollution blocked these rays out, and that altitude was good for health. In subsequent articles over the next five years, "How to Keep Well" offered pronouncements on the benefits of sunlight and reported insights on the matter from the medical profession.[2]

In 1921, Evans tempered his enthusiasm, cautioning that a more thorough understanding of light was important and that, though the tendency was to encourage its use, few knew what they were doing. Doctors were right to worry about the still-hazy understanding of wavelengths. One day, Evans thought, "Smart Alecky young blades will be laughing at our shotgun prescriptions of light." By 1923, he had even surpassed his former confidence, asserting, "If nature has a panacea for human ills and prevention of every human ill, a universal remedy— that potent draft is sunlight." Although Evans, writing excitedly of Saleeby's *Sunlight and Health*, mentioned that treatment might, under certain circumstances, do harm, such caveats were easy to dismiss alongside words like "panacea."[3]

In later years, Evans would mirror the medical establishment's ambivalence, combining a more circumspect regard for sunlight treatments with an attempt to educate the public about the nature of natural light and its limits. He wrote repeatedly about the topics that doctors argued mattered: the importance of careful dosage, the problems of polluted atmosphere, and the limitations of indoor sunlight. He tried to keep his public informed about glass.[4]

And Evans's was not the only voice. In Los Angeles, *Times* columnist Lulu Hunt Peters told of the miracles sunlight performed. Though she did not describe it as a panacea, she did tout the curative effects of sunlight therapy on dry and scaly skin, asthma, tuberculosis, bladder conditions, runny ears, polio, and of course, rickets. In admonishing readers not to lay too much faith in the mercury vapor lamp, she reminded them that "sunlight alone is a most valuable curative agent, and the mother and baby should have all the sunlight possible."[5]

Peters and Evans expressed profound medical hopes, but really, their story was only the tip of the sunlight-is-healthy iceberg. The battle against darkness was not simply a campaign for a well America. In their assertions about a broad role for light, these authorities and others sought a heritage for sunlit wellness much older than the nation itself. They turned to classical civilizations, which, the thinking went, had grown strong because the sun was on their side.

Books describing ultraviolet or sunlight therapy traced a heritage back to the Aztecs, Egyptians, Incas, Arabians, Chaldeans, Persians, Israelites, and Philistines.[6] Their most frequent ancient sources of sunlight thinking, however, were the Romans and, even more, the Greeks, whose worship of Apollo evinced a deep faith that the sun could do wonderful things, and whose physicians, led by the father of modern medicine, Hippocrates, had uncovered the wellspring of health. According to one author, these pioneers understood light's benefits so well that they employed it in treating conditions like "muscle wasting," and the doctor Aesculapius had made it a part of a "general physical culture" regimen.[7]

For other advocates, it was necessary to move back beyond history and civilization to fully understand the lineage of sun thinking. In many of his works, Luckiesh wrote of primitives whose gut feeling of its benefits had led them to worship the sun. In time, though, the "primitive being who seized a flaming fagot from the open campfire and carried it into his cave laid the foundation of the modern home." This act of protocivilization was a great moment in world history, but it also set a dangerous precedent. Fortunately, the savage's move from the open to the cave was only partial, as he would have been unable to survive a completely unlit experience: "With the sun no longer acting directly upon his body or wound or scanty apparel, the savage would be in grave danger from germs. Lacking in knowledge and inclination, this savage race would likely decay and even perish if it withdrew from the sun to the extent that most civilized persons have."[8]

There was far less consensus about what happened to the accumulated knowledge of primitive and ancient sun worshippers. Some authorities simply sidestepped the problem. Others argued that a rising faith in surgery had chased more naturalistic alternatives from popular use. But the most common explanation held that Christians purged sunlight therapy from medical practice in the Middle Ages when they decided it was paganism. Robert Aitken, a doctor and lecturer on skin diseases at Edinburgh University, was among oversensitive Christianity's most aggressive critics:

> This cult of the sun seems to have been looked upon as pagan, and with the advent of Christianity there came a revulsion of feeling, and the practice of sunworship fell into abeyance. The pendulum swung too far, and in addition to the decline of sun-worship there seems to have been a lack of recognition of its beneficial properties, and it is only comparatively recently that the benefits of sunshine have again been realized.[9]

The group most actively attempting to correct this excessive conservatism identified themselves as modern sun worshippers and looked to the past for guidance. In his introduction to the English translation of Rollier's *Heliotherapy*, Sir Henry Gauvain called Leysin's founder the high priest of the cult. For De Kruif, he was a prophet. In an ad for *Hygeia*, the drug company Merck claimed that pharmacists, physicians, and health authorities were united in an attempt to improve the public's welfare. The ad, titled "Ancient and Modern Sun Worship," showed a figure bathed in sunlight and related an interaction in which Diogenes asked Alexander the Great about the empire's needs, to which Alexander answered only, "step aside" and "do not shut out my light." The new sun worship had been reconstituted, and a pharmaceutical company, far from the standard devotee of such movements, was apparently a believer.[10]

Those who wrote about sun worship would use religious terms, but they did not manifest anything like a well-developed faith. And it would be a mistake to think of a drug company as a new church or to assume people in 1930 prayed at a Merck altar. Still, it makes no more sense to dismiss these half-winking supplicants. They were not foolish, and they were not halfhearted in their pursuit of mental and physical wellbeing. In the *Nation*, Stuart Chase told the story of his indoctrination into a group of a hundred naked men, all of them colored like "South Sea Islanders." He found his friends and fellow worshippers at Boston's L Street Bathhouse. In time, Chase would learn what he called the dogmas of his cult and would come to revere its "high priest," Richards. Stripped of all the encumberments of civilization, members of the group gained wisdom and shed nervousness and haste. The requirements for admission to this bunch were few, but a man free from the clock faired best because he had the right temperament and the requisite free time. Chase's overwhelming sense of freedom made him feel like a new animal; he wondered, "Who shall say of what strange and primitive juices, what fantastic combinations of electrons, the true sun-worshiper is made?"[11]

Newspapers and magazines joined the medical experts and Stuart Chase in their celebration. According to the *Chicago Daily Tribune*, "If we've become sun worshipers, it's a fine universal religion. But nothing new. There have always been sun worshipers in the true religious sense of the word." As the *New York Times*, which elsewhere looked favorably on Chase's article, explained, "The modern sun cult has attained almost the proportions of a new religion." *Electrical World* suggested that humanity needed a little more artificial sun worship, and *Popular Mechanics* pointed out the revival of "one of the world's oldest religions, sun worship." There were few places that the faith did not reach, a fact *Nation's Health* realized when it noted that "various forms of religion and social practices are extant even among highly civilized groups in the present day in which some element of sun worship plays a prominent part."[12]

But there was often another layer to the sun cult that had broader objectives and deeper concerns. Some contended that it was wrong to focus too closely on classical civilization or primitive peoples. Darkness was a troubling problem for a truly modern nation. Often uniting—or perhaps confusing—nation and race, these thinkers claimed that white people must gather strength from sunlight or civilization was in peril. *Delineator*, one of the foremost women's magazines, mixed together a host of issues when it wrote, "Most rachitic children are short, and dwarfism may be produced. Thus the efficiency of the working power of a nation may be distinctly affected, a fact that was recognized in the

Great War." The confounding of child rearing, work, and war made the subtexts of John Howland's article hard to decode, but its overall sentiment was unambiguous. Darkness was a critical danger; it would help determine the fate of the nation.[13]

In 1929, Edgar Lloyd Hampton told his *Los Angeles Times* readers how California athletes had become the nation's greatest. Their advantage started young; at age three and a half, children were as mature as the average five-year-old; by four, they were two to three years ahead of their peers. These advantages carried through their teen years, with college freshman considerably larger than first-years from other states. Hampton believed that no genetic explanation could suffice. California's healthful sunshine and fresh, vitamin-rich fruits had made its children strong. In another article, the story was the same; USC athletes had dominated an athletic competition—the reason: "The Trojans had in their hearts and muscles just a little more solidified sunshine."[14]

With popular wisdom confirming that sunlight was necessary for the construction of healthy bodies, Howland felt confident that he was right. But he contributed only a small part of a large and disturbing story. For some of his contemporaries, the difference between Los Angeles and New York was not the primary issue. Germany was, for them, the world's heliotherapy leader, its physical culture unmatched. In its article "Germany's Quest of Greater Strength and Beauty," the *Washington Post* pointed out the profound pull sun worship had there: an enthusiastic public traipsed around beaches naked and basked in the sun's ultraviolet rays. While noting that the medical treatment was not for the feeble and had to be dispensed along scientific lines, the article, especially its title, sounded an alarm: sun therapy could create a more potent people, and falling behind the trend, the United States could become a country of relative weaklings. Muscovites, too, were building bodies strong and using sunlight as a tool to overcome their poor athletic facilities:

> They take sun baths on balconies, in yards or on roofs, while along the Moscow River the absence of an ethical code as to costume permits of the fullest benefit from ultra-violet rays. Those fortunate enough to spend their vacations in the Crimea or on the Black Sea return the blackest of all, to the great envy of their friends.

Good communists, their goals were twofold, a healthy race and a cooperative spirit.[15]

Nation and race crept into this sunshine discourse in curious ways as medical and public health professionals ushered in a new wave of eugenic thinking. This revision inserted environmental concerns—sometimes awkwardly—into the conventional theory. Caleb Saleeby, chairman of Britain's Sunlight League and Birthrate Commission, was a leader of this theoretical turn. His 1921 book, *The Eugenic Prospect*, which a *New York Times* review praised, argued that genetics merely indicated the possible life outcomes that environment could bring about. Basically, genetics limited humans but did not determine what they could be. This view, in which potential played a profound role, asked how best to stack the deck in favor of a lively people. Saleeby confessed that he was jealous of American hardiness. A trip to Liverpool revealed a "pitiful little crowd of hoarse, rickety, stunted anaemic children." The problems in Britain, as elsewhere, were the demands of modern civilization and big, gloomy cities, which had torn people from traditional and healthy habits.[16]

But Saleeby was convinced the future did not have to be bleak. America, for example, had skyscrapers that rose above smoke and regulations that chased it away altogether. Even Pittsburgh, the icon of sooty skies, had overcome its problems, joining Chicago, Washington, and Philadelphia as successes so clear that walking on a Saturday evening, Saleeby "could see no offence whatever to the eye of racial hygiene." Often, the definitions of race and nation blurred for the Briton, but his concerns were great and consistent:

> Bodily and mental vigour, superabundant energy, daring, achieving, pioneering, breaking new ground, actuated by "divine discontent," fearing nought— these are generally whether in the individual or the nation, the products of Health, dependent with rare exceptions upon good parenthood, childhood loved, understood, well-nourished, and youth guarded and disciplined.[17]

Saleeby was sounding the clarion call for "heliohygiene," a distinct and important redirection of the older heliotherapy movement. In *Sunlight and Health*, he offered thinking that resonated well with Luckiesh's work-and-vision model. Having evolved in light, the question for modern sun worshippers had to be, "What kind of men, women and children do they produce?" Saleeby's earlier celebration of the United States gave way to praise of Canada as the ideal, a place where sun and cold worked unparalleled wonders for urban children. But location was a secondary point, as no matter where things had gone right, the stakes were high everywhere. Throughout history, civilizations had suffered eventual decline, and England was on its way. At the First International Conference on Light, the eugenist's paper "From Heliotherapy to Heliohygiene" called

for Rollier's recognition with a Nobel Prize in medicine and argued vehemently
for a healthy outdoor lifestyle: "We must restore pure air and light to our cities,
or surrender the leadership of mankind to other nations, which are now building cities where pure water, clean air, unsullied sunlight and uncontaminated
food will be the common heritage of all their children."[18]

Saleeby's theories, called positive eugenics by its adherents—for their emphasis on making good races better rather than breeding out the bad ones—
resonated with American thinking. In 1928, the Third Race Betterment Conference convened in Battle Creek, Michigan. According to Race Betterment
Foundation president John Harvey Kellogg, it was imperative to counteract de-
generative influences, "to bring together, to unify and to stimulate the activities
that are working for eugenics and personal hygiene and a practical application
of physiology and biology to human living."[19]

Conference and University of Michigan president C. C. Little asked a question that must have plagued eugenists and heliohygienists, "Shall We Live Longer
and Should We?" Medical improvements promised to prolong the lives of inferior types, permitting the perpetuation of mental and physical defects and potentially beginning racial degeneration. Nevertheless, Little maintained, there was
much to be hopeful about, for the difficulties facing modern humans were also
breeding tougher people:

> We are the hardiest specimens of the human race that could possibly be de-
> vised; for if living in bad air and without sunlight and without proper nutrition
> and with no knowledge whatever of the use of clothing or the lack of it, if all the
> various eccentricities and idiosyncrasies that we have developed are any crite-
> rion at all, then certainly we have done wonders to survive, and we represent, if
> not the most beautiful physical derivatives of the human race, at least, I think,
> the thickest-shelled variety.

In time, when selection pressures became harsher and the least fit had to care
for themselves, they would die out, and the most capable—made even more
capable by the vicissitudes of modern life—would again thrive. Though the implications of Little's brief comment on sunlight might be read as a celebration of
urban darkness—the benevolent selector that would eventually make people
better—it is unlikely that that was his intent. Sunlight was a good, and its opposite was the enemy. This was not the end of talk about such matters. The conference proceedings offered a section on the benefits of sunlight therapy. In one of
those papers, MIT biochemist and physiologist John W. M. Bunker considered

the future of the "scientific application of light to life, health and happiness," pondering whether it could "be an aid to the development of a happier and a better behaved race"[20]

There was another, more sinister line of eugenic thinking promising a better future: specific knowledge about darkness would help selectively breed a racially superior nation. According to Benjamin Goldberg, it was a known fact that certain races were better suited to certain environments. Goldberg argued that although tuberculosis was thriving in cities where small spaces packed people into dark rooms that bred disease, not all those exposed suffered similarly. White people, already accustomed to civilization, had built up immunity, while black people, Mexicans, Puerto Ricans, and Filipinos—the "primitive racial groups"—ill-suited to urban life and usually forced to endure the worst conditions, suffered immeasurably. Disease plagued these lesser sorts.[21]

Goldberg and others argued that different races faired differently when it came to the diseases of darkness. In spite of their correct conclusion that extra skin pigment resulted in elevated rickets rates, most of expert theorizing was based more on racist assumptions than accurate science. Authorities argued that the races best suited for powerful sunshine developed the diseases of darkness at a staggering rate when far from their natural environments. Black people suffered most severely, but Italians faced grave dangers too. Dark skin in a dark city was no better than fairness in bright sunlight.

Such arguments gave credence to Luckiesh's racial theory of rickets and population distribution. His 1926 book, coauthored with Pacini, argued that, millennia ago, lighter-skinned people fled equatorial regions, unable to stand bright light. In the tougher—darker and colder—climates, "slaughter of the unfit took place at a fast rate." It required ingenuity to survive. No such selection pressures existed where darker-skinned people lived, explaining why southern Europeans and Africans were less intelligent and resourceful. "Half breeds" were worst off. Suited to no climate in particular, they had little hope anywhere. In American Anthropologist, Frederick G. Murray offered a similar developmental model but linked it more clearly to rickets. Fair and blond people could not thrive in too much light and heat, making them unsuited to the south. Darker-skinned people were the opposite, dying out in northern environments as a result of unsustainably high rickets rates. Saleeby predicted a similar fate for all populations that endured unsuitably dark places, because the "rickety contraction and distortion of the female pelvis involves the gravest risks to motherhood and the maintenance of the race." Climatic maladaptations could carry grave consequences. All of these arguments taken together gave voice to a unitary fear that people were built for life in particular sunlit climates. Outside of these parameters, the various races, however eugenists wanted to define them, suffered.[22]

With the celebration of sunlight and tanning, a curious thing happened to ideals of skin color. In Anglo-American culture, pale skin had traditionally been the standard: it connoted beauty and status, a life free from toil outdoors. But that began to change around the turn of the century. By 1930, a new aesthetic ideal was in place, with men, women, and children heading out into the sun. Some counseled against the goal of "blackest" skin to which Muscovites aspired. According to the *Chicago Daily Tribune*'s author Gladys Huntington Bevans, the goal was "sun tan or sun brown, but not deep bronze," for too much pigment could be dangerous. Elsewhere, there was no such distinction: after weeks in the sun, Rollier's pitiful, runty wrecks turned "a deep mahogany brown," transformed into "brown-skinned youngsters," "bronzed aborigines." People needed "plenty of sunshine" to excise pale listlessness, and in Los Angeles they varied from the "color of weak cocoa to a deep, rich chocolate suggestive of African shores."[23]

Some eugenist minds might have been disturbed by the new brown-is-beautiful movement. Indeed, there was much in America's history that had conditioned the opposite thinking, but the evolutionary theories offered a satisfying solution. There was a difference between tanned skin and hereditary dark skin. The latter was unhealthy and dangerous in northern regions, evidence that black people were unsuited to America's temperate environment; the former represented a spectacular adaptation to that environment, which carried—and connoted—health. Built into these arguments about skin color was a way of avoiding the unpleasant conclusion that inferior, dark races had an inroad to beauty and wellness.[24]

In reconfiguring the sunlight debate along sanitary, constitutional, and hygienic grounds, experts moved light therapy into public view. Tanning was not a matter for the sick only. The sun was there for all to worship, and that practice, if done correctly, was worthwhile. How to supplicate was an important topic, but first people needed homes fit for devotees. By 1930, with glass occluding the best parts of natural light and big cities continuing to grow, the nation endured darkness problems far larger than planners in the nineteen teens had envisioned. Armed with hope and knowledge, a new group of builders decided it might be worthwhile to try constructing for sunlight again. Not everyone, however, had benevolent motives. Some wanted communities in which all got their ration of health, but far more were interested primarily in profits.

At the Tenth National Conference on Housing in Philadelphia, two present-
ers argued in favor of more attention to sunshine in housing. Both concluded
that better-lit homes were necessary for health and that conventional electric
bulbs were not the answer. In his paper "The Health-Giving Ultraviolet Rays,"
Donald C. Stockbarger told his audience that it need not follow the letter of doctors' admonitions to "get out into the sunshine." Glass innovations offered an
intriguing—though not ideal—alternative; while a stroll outside might be nice,
special, transmitting windows brought similar health. Architects had a responsibility to design apartments with that in mind, ensuring that the rooms most frequently occupied were the best lit. Stockbarger concluded his comments with
his version of the old adage of not looking a gift horse in the mouth:

> If you were to purchase a parcel of land on which very pure spring water
> flowed, would you not consider the location of a house with this in mind? If
> natural gas were available, would you not immediately plan to make use of it for
> heat and power? Then why not utilize as much as possible of the ultraviolet in
> the sunshine which is available on nearly every parcel of land.[25]

A remarkable French design one-upped Stockbarger: the revolving solarium
at a health resort in Aix-les-Bains chased its gift horse down. The facility followed the sun's movement across the sky and added mirrors for more constant
tanning. The New York Times reported on similar plans displayed in Paris by the
French Society of Professional Architects and so did F. W. Parsons in the article
"Sun Worship," which celebrated sunlight innovations. That article also mentioned George Bernard Shaw's new three-sided revolving house, a "clever sun
trap" with specially glazed ultraviolet-transmitting windows.[26]

While such designs probably were not feasible on a large scale, Literary Digest reported on another promising French innovation that it believed might
have wider application. The Digest article reminded readers of the willing alle-
giance between doctors and druggists in spreading the "Crusade of Light." Pre-
pared to do their part, architects were now "disemboweling our facades, opening
huge bays [and] they are even building houses of glass." But tenants occupy-
ing the interiors of large buildings needed an alternative, or they would have to
"vegetate in darkness"—the solution, a system of motorized mirrors that, for a
cost of $3,250, could light a six-story building with sunshine. Perhaps this idea
helped inspire General Electric's installation on Central Park South in Manhattan, where the consulting engineer painted the courtyard white and installed
timer-controlled, thousand-watt bulbs on the ninth floor to bring sunshine to

everyone below. According to one glowing trade-magazine report, technology was doing great things: indoor lights tanned, air conditioners simulated autumn breezes, and the GE courtyard design employed a lighting wonder that, despite a lack of ultraviolet rays, seemed almost natural: "[It] rises and sets in a fashion that to some extent duplicates the lighting conditions created by the real sun." It was little wonder, the article continued, that residents of the building's upper floors were considering moving down; while the real sun might fail to shine through clouds, these bulbs were always reliable.[27]

Most of the solutions to darkness were less elaborate than Aix-les-Bains's revolving solarium or 100 Central Park South. In a collaborative project with *Scientific American* editor E. E. Free, Ronald Millar told of people who built penthouses in order to lie out in the sun "in the hope of getting a healthful and fashionable coat of tan." Though Millar was unconvinced that homes on the roof were a viable solution, as pollution obscured ultraviolet light, dirty skies did not stop the countless many who, the *New York Times* reported, were heading onto their roofs in search of health. One author called his steel-and-glass penthouse apartment, open on four sides to the sun, a "figurative" tent.[28]

Written in 1924, when the hope of a glass structure under the sun that offered access to the whole spectrum of light was still alive, the *Times* author's tent might have seemed worthwhile. Seven years later, 500 Fifth Avenue had no excuse for ignoring the transmission properties of its windows. Nevertheless, it advertised its luxurious location without attention to the vital health element of ultraviolet light. Its exposures on Fifth Avenue, the library, and Bryant Park permitted the building to boast sunshine for 85 percent of its windows. Towers like 500 Fifth Avenue rose up from city blocks in order to secure light, and building designers offered innovative architectural solutions to meet new needs. Sometime before, factories had begun using sawtooth skylights, which were ridged for maximum natural light. In 1929 and 1930, *Architectural Record* offered a tower-in-the-park equivalent. The design employed outer walls with multiple vertical peaks and valleys; more angles enabled glaziers to coat more of each room with glass. Other authors wanted planners to pay closer attention to the sites of their structures. Sun charts updated Swan and Tuttle's work on urban sunlight and shadow, attempting visual representations of the path the sun took as it moved through the sky over a given location. That was just the type of rigor that housing-reform advocate Henry Wright called for in his popular and highly praised book *Rehousing Urban America*. In a short section, he called for new studies of the proper orientation of buildings. Though little explored, the principles necessary to secure adequate fresh air and ventilation would probably come easy, but that was not the case with sunshine. Still, it was time for this embryonic field to grow up; conventional architecture did not secure "sunlight of any intensity or of any value for therapeutic purposes or for the elimination of bacteria."[29]

Newspaper advertisements began writing of the many homes that provided a sunny room or two, a solarium, or a sun parlor. For Gotham's wealthy, fancy apartment buildings often offered these spaces. London Terrace, on Twenty-Third Street, counseled prospective tenants, "Sunshine is good for you"; located near the Hudson, its new Marine Deck allowed bathers to absorb ultraviolet light to their "hearts content," and a children's sun terrace offered parents an excellent opportunity to keep their youngsters healthy. The Century, at 25 Central Park West, celebrated its three-, four-, and six-room "solarium apartments" and its private terraces and windows, which admitted the ultraviolet spectrum. The Majestic Apartments, at Central Park West and Seventy-Second Street boasted the installation of Vita Glass. The San Carlos did even better; it was the first hotel apartment to use only Vita Glass throughout: "Thus the vital rays that contribute so much to health can penetrate freely into each room to help children or adults to sturdy bodies and better health."[30]

The new glassmakers were offering their products as good alternatives to inferior, traditional options in cities. They claimed that they could bring at least a minimum of bright healthfulness to shut-in, big-city workers and residents, but that does not mean they were willing to cede the suburban market. Although it was outside the city that play outdoors was easiest and products like Vita Glass, reason might have dictated, were most dispensable, it was also there that the greatest limits of glass—shadows and pollution—had been overcome. With decentralization came less smoke and therefore better, more robust light. Commercial suburban endeavors, eager to sell themselves as cutting edge, did not miss the opportunity. In 1929, the New York Times estimated that a smart and enterprising family could build an English cottage fully equipped with health glass near to the city for $12,000. It would permit sunbathing even on winter days, and with sunrooms almost as large as bedrooms, lying out became comfortable. Dating from the nineteen teens, Forest Hills was a planned community that featured row houses and single-family dwellings with large porches, some identified as "sunporches" on the architectural drawings. By 1930, the community offered another boon for sunshine seekers: Vita Glass lounges. This was not unique. Some of Scarsdale Downs's master baths had three walls and an entire ceiling of Vita Glass, so that families "could acquire a healthy sun-tan in this gorgeous room all through the winter." As Vita Glass sales figures show, it was not an overwhelming success in any market, but suburban developers seemed willing to try it to get a leg up on the competition.[31]

In advertisements featuring lovely suburban scenes or well-spaced tract housing, glass innovators and realtors did their best to reinforce the association of suburbia with modern, sunlit construction. One *Chicago Daily Tribune* Vita Glass ad told of sophisticated urbanites and builders who had already improved the glass in their homes, overcoming a fundamental danger of this "city chained age." To suburban doubters, Vita Glass would send a list of well-glazed apartments; for those ready to catch up with their metropolitan neighbors, it was time to contact one of the many retailers selling the product and secure summertime health all year. Now was the time for suburban homeowners to join modern apartment residents, to glaze with Vita Glass, and to reject the apparent inevitability of "seasonal sun-starvation."[32]

Vita Glass apartments in cities and suburban homes were not the only choices for sun-sensitive homeowners. A handful of builders began arguing that, just as workplaces and schools could provide the benefits of the out-of-doors without leaving the indoors, homes with improved lighting could outdo the sun. With air conditioning and modern illumination, humanity would return to the cave. A couple of *Los Angeles Times* authors even predicted a burrowing boom: builders might stop constructing skyward altogether, opting instead to tunnel underground habitations.[33]

None of these arguments, however, sought a broadly accessible solution to the dark-housing problem. Wright's ultimate goal was better living for all, but his building-orientation argument was not necessarily about caring for the needy. The earlier tenement reform movement had encouraged the city to look at individual success stories—in which some people enjoyed great advantages while the majority suffered monstrous cruelties—as unacceptable. In revisiting its planning legislation, a fresh group of New York reformers began to wonder if the city had done enough. The mayor's Subcommittee on Housing, Zoning, and Distribution of Population claimed that two million people lived under inadequate conditions. Poor natural light was clear evidence of that inadequacy—a sad fact, the *New York Times* reported, "[since] a ray of light travels 93,000,000 miles to reach us from the sun. Then when it comes here we deliberately throw away and exclude its most valuable qualities." After study, the subcommittee concluded that the city's height and setback rules were too lax. Chicagoans were having similar debates: their politicians deliberated over a proposed amendment to the zoning law that would have permitted tall structures without setbacks. Coroner Herman Bundesen—later health commissioner—was displeased; he claimed that the resulting decrease in light would be disastrous. "Isolation, light, and air," he maintained, were "all vital forces in combating communicable diseases."[34]

In the late 1930s, the Work Projects Administration produced a number of housing posters for New York that looked to drum up support for improved residences. One read "Planned Housing Fights Disease." The message was unambiguous, its picture simple: a blue background and a bright yellow sun shooting rays straight through a microbe. Ten years earlier, Lawrence Veiller had lamented that planned housing did not do enough disease fighting. Always a champion of restrictive legislation and not fully convinced that municipal construction could work, he spearheaded an attempt to reform New York's housing system to secure more sunlight. For Veiller, New York was stuck "in the dark ages, compared with other large cities." He proposed a new study of building designs no more than two rooms deep and an act that would require the use of ultraviolet-porous glass in new apartments. Vita Glass was pleased that Veiller had singled out its product as one of humanity's hopes. It proudly advertised the housing expert's endorsement, telling readers that he believed it was within their means to reglaze their homes. But Veiller saw other options too; his journal, *Housing*, favorably reviewed Luckiesh's *Artificial Sunlight*, writing enthusiastically of the engineer's work studying sunshine substitutes. The endorsement could not say whether humanity would ever find "satisfactory substitutes for God's sunlight and fresh air," but it was certain that the solutions discussed in the manuscript showed great promise. Planners and "housing practitioners" would benefit from its insights when it came to artificial and natural daylight.[35]

Catherine Bauer agreed with Veiller that the time had come for better planning, but she articulated a vision far beyond the simple minimum standards that the old reformer and his cohort had advocated. For her, zoning had not done much to undermine the problematic laissez-faire paradigm. Cities were "enmeshed in a spiral of congestion, high land-prices, skyscrapers, worse congestion, higher land-prices and higher skyscrapers." The marginal improvement tenement laws affected were entirely too marginal for an industrial poor housed in dark hovels devoid of "a breath of fresh air or a direct ray of sun." For Bauer, this was only part of the problem. All homes in cities, "for rich and less rich alike," were unworkable for, among other reasons, "fundamental deficiencies in light." But she was an optimist. Housing's time had come, and the public was realizing that it would have to build. New construction would not be the "congested tenements, wooden three- and four-deckers, and jerry built, jostling bungalows" of the previous ten years. Improved designs would have higher

Figure 4.1. "Planned Housing Fights Disease," late 1930s. By the People, for the People: Posters from the WPA, Work Projects Administration Poster Collection, Prints and Photographs Collection, Library of Congress. Courtesy of the Library of Congress.

standards, securing for residents a new minimum of light with balconies and roof terraces.[36]

Lewis Mumford also saw a problem with the twentieth-century American city. Like Bauer, he thought that communities had to get involved in housing and that regional planning was critical, but his writing betrayed a deeper pessimism about substantial change and a more profound concern about light. Mumford acknowledged that cities were natural, like "a cave, a run of mackerel or an ant-heap." People would inevitably cluster, but culture and technology could determine what these places were like. For Mumford, the ideal city would foster civic spirit in a healthful environment. Modern cities were bloated, inhuman, and chaotic masses, overcrowded and stifled by pollution. He wanted to reintroduce order with a new urban form. His towns, unlike the atomized suburb, would have downtown civic centers and would balance "home, industry, and market, between political, social, and recreational functions." Put simply, Mumford hoped for "regional centers" that "would combine the hygienic advantages of the open suburbs with the social advantages of the big city."[37]

A more natural lifestyle was part of Mumford's new, humanized vision. He wrote about the rebirth of Kepler's sun cult and cited researchers like Downes, Blunt, and Pasteur who had helped reveal that "indifference to darkness and dirt [was] a monstrous barbarism." Rickets findings, tuberculosis data, and sanitation studies had proven that the old, polluted city, with damp and dark factories and slum homes provided "ideal conditions for breeding bacteria." Living organisms required a life-sustaining environment, but "modern hygiene" had shown, "biologically speaking," twentieth-century cities were "life-inimical or life-destructive environments." Perhaps these hulking metropolises could remain, but they would have to evolve drastically, and effective change would be most difficult. Big buildings were nearly impossible to construct on sound hygienic principles, and cities with pollution-belching factories could never bring to the citizenry what they needed. As for the preexisting reform model, that was good for little; zoning for health was fine, but New York's law, for example, failed to secure adequate minimum standards.[38]

Bauer and Mumford would never get their wish of well-planned public housing for a significant proportion of the population. New York, however, may be the closest Americans came to this goal. Though it too suffered from a public housing shortage, the city built more units than any other, and its early plans to provide healthful homes to its neediest citizens showed a concern for sunlight. In 1935, at the very beginning of its construction efforts, Langdon Post, chairman of the City Housing Authority, called for "ample living quarters and at least a minimum of modern sanitary conveniences, plenty of sunlight and fresh air." That same year, a group of ten well-known architects unveiled a model project that did just that.[39]

Public officials in New York stood up and spoke out about the smart design that made their projects special—at the laying of cornerstones and celebrations to open the Williamsburg (with 1,630 apartments) and Red Hook projects (2,545) and the East River (1,170) and Vladeck Houses (1,531). In the case of Queensbridge (3,149), the city consulted Henry N. Wright, son of the famous housing advocate. Considering unavoidable practical constraints, he wrote, the plans for Queensbridge were good. Well-spaced buildings, one room deep, offered ideal sunshine conditions, but this plan, which permitted north-south exposure, would secure considerable light all year, and the Y-shaped buildings were a drastic improvement over X-design plans at other locations.[40]

It is far harder to gauge the commitment of private developers to sunlight, but the Federal Housing Administration (FHA) operated at the intersection of public and private. The government program insured mortgages and essentially remade the housing market in the 1930s. Previously, lending terms made it difficult for many Americans to buy houses. The FHA changed all of that by backing mortgages to facilitate borrower-friendly conditions. This was no government handout, however. The FHA only wanted to insure sound investments and, as a result, held its wallet tight and looked to avoid risky bets.

The FHA had lots of eligibility requirements for properties, and some of them pertained to windows and light. Each room needed at least one window "having not less than one square foot of glass for each ten square feet of floor area." Appraisers had to make sure that a neighboring building did not tower too high over or stand too close to a required window. Local FHA requirements could be more restrictive; nearly all prohibited mandated windows from opening onto interior courts.[41]

The effect of the FHA on the housing market is too large to measure. Between 1935 (its first year of existence) and 1938, 410,000 sales received FHA insurance, but that is not the full extent of its reach. Nobody knows how many developers planned with the program in mind, and by setting minimum standards for so many homes, it influenced construction on yet more projects throughout the nation.[42]

There are two ways to make sense of this increased government commitment to better designed homes. Congress charged the FHA with a mission to "encourage improvement in housing standards and conditions." But the idea was not just to promote better housing; the FHA program was insurance, and insurance is risk averse. The FHA recognized that it could make choices "entirely apart from the factors involved in the safety of the investment itself," but it would not. The program was convinced that "the qualities which produce a satisfactory social condition also tend to assure economic soundness." In short, one way to read the FHA commitment to sunlight is that it saw social benefits in sunlight; another is that bright homes were solid investments because good light paid.[43]

When it came to the shape of a house or the height of a building, planners asserted their role and urban critics found their voices, but both groups knew that their effects on the lives of residents could never be certain. Better buildings *could* confer innumerable advantages to long-suffering residents; whether they *did* would be decided when construction stopped and the front door slammed. Fortunately, other concerned citizens were taking on that problem, convinced that their ability to teach women how to run their homes and raise their children was most important. Advice came from all sides, with doctors, officials from government bureaus, nutrition experts, and authors in parenting magazines looking to make a difference.

The Department of Labor Children's Bureau made its case for better parenting often but never more comprehensively than in its book-length educational tracts *Prenatal Care, Infant Care,* and *The Child from One to Six.* The bureau pointed to evidence that the needs of children started even before birth. Expectant mothers were supposed to spend a couple of hours per day in the sunshine. That exposure would ensure that children were born with the building blocks for good teeth and bones. After birth, however, sunlight needs grew. Outdoor sunbaths in spring and summer could begin when babies were just three or four weeks old. The common aversion to putting babies outside in wintertime was overstated, as the temperature in the sun averaged 40–50 degrees warmer than the thermometer read. If parents did not have a sunporch, they could substitute an open window. That would also help in the case of exceptionally cold weather when a heated room with an open window could allow the healthy sun in while mitigating a winter freeze. The initial duration of treatment had to vary according to the little subject's skin color. Exposure increased over time as children became tanned and, therefore, less photosensitive. For slightly older youngsters, daily play in the sun was a necessity, helping them grow strong bones and make proper biochemical use of their food.[44]

The bureau's *Sunlight for Babies* articulated a clearer, more focused argument. In 1926, the pamphlet, a revision of an article published in the *American Physical Education Review* the year before, appeared with a smiling youngster in a highchair on its cover. The text told parents of the division of light into a "well-known spectrum of colors" and lesser-known invisible regions. It cautioned that winter sunlight was impoverished, that the temperate zone did not give a full spectrum, and that bundling a baby might do harm. Five years later, in a new revised version, much of the content was the same, but there was far less emphasis on the nature of sunlight, perhaps because the public already knew about it, perhaps to save space. The focus was almost exclusively on rickets and the complications it brought, from attendant diseases to the life-long deformity of girls' pelvic bones that had so worried eugenists. By 1931, moreover, the means of prevention had changed. In a development that presaged a later, more extreme shift, the pamphlet elevated cod liver oil to a more prominent place.[45]

According to an October 1930 letter from bureau chief Grace Abbott, the agency had reason to be proud about its public education campaign. In that year alone, 63,344 copies of *Sunlight for Babies* had been distributed, second in its twelve-pamphlet series only to *Why Sleep?* It appears that mothers got the message. In 1932, Mrs. W. T. McDerran wrote to the Children's Bureau for clarification. She had read *Sunlight for Babies* and noted that it counseled her to keep her youngster out of the sun between the hours of eleven and three. Her physician, however, had claimed that the hours between ten and two, when ultraviolet radiation was strongest, were critical. The bureau's response told Mrs. McDerran that the extra ultraviolet light was more than offset by harsh heat, which permitted only short-duration tanning. It would be better to let her child out in the morning and late afternoon.[46]

Letters to the Children's Bureau varied considerably. They came from, among other places, Central City, Nebraska; Columbus, Ohio; Chicago; Philadelphia; Gurdon, Arkansas; and Balboa Heights in the Canal Zone. Parents sought counsel on a variety of subjects, but most frequently, they wanted to know whether a sunbath would help their youngsters. Generally, the bureau sent a copy of *Sunlight for Babies* with the caveat that doctors knew best. Some questions indicated that parents were interested in the role artificial sunlamps could play in their children's health. One concerned father from San Antonio wondered if such a product would help his little one fight off colds, a curious question, the bureau told him, because Texas had abundant natural sun. Another parent wrote that her child's doctor had counseled a $2.00-per-treatment sunbath regimen. Unsurprisingly, a lamp salesman said that was unnecessary: home treatments were sufficient and economical. A consistent advocate for professional supervision, the bureau did not agree. Specific queries about lamp effectiveness were somewhat common, but responses were generally noncommittal, encouraging concerned parents to seek more information from a doctor.[47]

Sun Babies, a film produced for the Children's Bureau, substitutes sad footage of rachitic children for the cute cartoons dotting *Sunlight for Babies*. The silent movie, with subtitles in Spanish and English, opens with children bathing

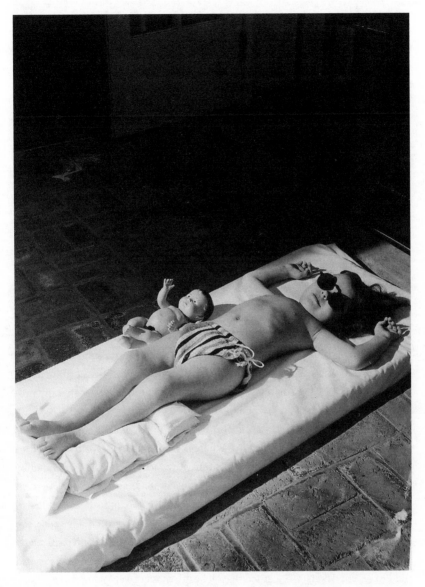

Figure 4.2. Small sunbather, ca. 1930s. Photo no. 102-G-116-3, Records of the Children's Bureau, United States Department of Labor, National Archives.

at the beach. The next shot, "sunshine in the city" shows youngsters playing on fire escapes and lying in the sun. It describes the problems of rickets for the development of youngsters, offers up-to-date information about the sunshine needs of and darkness dangers for children, and encourages tanning regimens or cod liver oil when necessary. The movie ends by telling parents, "By sea or on land— / The sun babies of today, promise healthy, happy men and women tomorrow."[48]

It was precisely this sort of work, reaching out to minorities and helping city kids, that the settlement houses claimed as their central charge. In New York City, these places, established to care for and assimilate immigrant communities, offered their wards sunlight. Christodora House, located on Avenue B and Ninth Street, in a community dominated by Russians, Germans, Poles, Italians, Irish, and Jews, hoped to educate immigrants, so that they might "grow into a better understanding of each other and their common city life." Its clubs were "the most obvious expression of this Americanizing influence." Many of the services the facility offered to its "needy" charges were medical, among them an ultraviolet lamp in the director's office. According to its reports from the mid-1920s, the Judson Health Center, in a mostly Italian district, sought to overcome its population's slavish obedience to old practices. Its program for educating parents and babies used "sun baths, natural or artificial according to the season." In its 1929 yearly report, Judson boasted a modest heliotherapy program, 199 patients had come for a total of 3,246 visits.[49]

Christodora House and the Judson Health Center offered small programs that brought sunlight to the slums, whereas the Mulberry Health Center attempted a robust public education venture. The facility's district, bounded by Canal, Houston, Bowery, and Broadway, housed 38,000, of whom 91 percent were foreign born or children of immigrants, mostly Italian. In 1927, Mulberry launched a two-year anti-rickets campaign. That sort of action made sense, the staff of the health center believed, because dark skin made the group particularly susceptible to the condition, one reason that nine in ten children in the area had mild to severe rickets. Concerned that all health matters were related—a weakly rachitic infant would be vulnerable to a host of illnesses—the goal of the activities was to reduce the incidence of pneumonia.[50]

This was not the first time that the settlement house's workers set out to improve diet and hygiene. A 1923 report claimed that 30 percent of children in the area suffered from malnutrition. It also recounted the considerable efforts underway to remedy that serious problem. A promising story told of a mother busy sunbathing her youngster when the neighborhood nutrition expert accidentally knocked. The parent informed her pleased—if surprised—visitor that she began the regimen after overhearing a similar program suggested to a neighbor. Other parents might have gotten information from materials like the Mulberry Health Center's calendar for 1925, which offered information on how to eat and live healthily. Alongside exhortations to drink milk, eat the right kinds of cereals, and consume plenty of fruit was the suggestion that children sleep with their windows open and "Play in the Sun to Keep the Doctor Away." These campaigns grew better directed and more extensive as information improved and concern about rickets deepened. The Mulberry Health Center began a districtwide traveling educational program in 1927. Nurses visited 685 mothers in the first year and successfully showed 491 how to sunbathe their children. Mulberry's yearly progress reports for 1929 and 1930 showed, respectively, 78 percent and 88 percent of mothers receiving similar education. In 1931, that percentage fell but the total number of children exposed to healthy tanning rose.[51]

Educational fliers and instructional pamphlets made sunlight and cod liver oil the two-part recipe for health. The short pamphlet *Rickets Must Go* pointed to a probable link between the childhood disease and anemia, colds and bronchitis, nervous-system deficiencies, and general debilities of the legs, knees, chest, and pelvis. That last problem had the added ill effect of a body poorly suited to childbirth. A shorter and more focused publication, *In Mulberry District*, connected pneumonia and rickets. Four babies, one with bent legs, appeared at the top of the page; below that, text told readers that a quarter of babies who died lost their lives to pneumonia. Below that were ten more pictures, nine of babies with bent legs and the information that 90 percent of babies had rickets, a disturbing statistic since rickets often led to pneumonia. While the suggested solution was cod liver oil, the message "prevent rickets and have less pneumonia" no doubt indicated to an audience sensitive to the light needs of their children that a bath in the bright sunshine was critical to their babies' health. Finally, in the leaflet *Sunshine Made the Difference in the Growth of These Chicks*, lessons learned on the farm traveled to the city. A graphic credited to the Wisconsin Agricultural Experimentation Station showed two silhouetted chickens, one hunched and small the other large and strong. They were the same age, fed and cared for in the same manner, but one spent half an hour in the sunshine every day. The moral was simple: "Try sunshine on your baby."[52]

The educational campaign to which Mulberry dedicated itself was not the only way that the public heard about the importance of sunlight. Serials offered similar evidence to an even broader audience, sometimes conveying the words of experts like the Children's Bureau, at other times providing conten-

tions all their own. Predictably, women, overwhelmingly the childcare provid-
ers and caretakers of the home, were often the targets of concerned counselors:
the Mulberry Health Center educated mothers about proper tanning, and the
Children's Bureau generally corresponded with women, though some men did
write. Throughout the popular press, there was a sense that women, not parents,
were the key to a healthy future. Paul De Kruif pushed the argument further in
"Sunshine, Open Air and Those Awful Colds in the Head," which appeared in
a 1928 issue of the Ladies' Home Journal: "And the modern mother, crude but
gorgeous experimenter that she's going to be with her babies has a grand chance
to see what the sun and open air will do to guard her youngsters' throats, noses
and chest; she's already baking them in the sun for the good of their bones."
It was the mother—for De Kruif, part caretaker part medical assistant—who
would bring sunlight to infants. This new item in her job description necessitated great vigilance. Another article agreed: prenatal care required sunbaths for
expectant mothers; and down the road, smart housewifery would necessitate
that those mothers open kitchen windows to ensure that their children's soon-
to-be eaten food was properly sanitized.[53]

While De Kruif celebrated maternal intuition and saw the woman of the
house as a pioneer, most articles in women's magazines were leaving nothing to
chance. They echoed the pronouncements in Sunlight for Babies, admonishing
mothers to be smart about their care. Sunshine was a tonic best taken outdoors
and without clothing, except for a head covering. It was dangerous to expose
too much of children's bodies too fast and important to make sure that the right
solar rays reached them. Tanning in the heat of the day could be dangerous, and
overexposure carried potentially catastrophic risks. Sunshine in cities was often
impoverished and winter light worth little, reasons that mothers should take
their youngsters out to the country or off to the beach. The language used in
these publications transformed mere motherhood into "modern" motherhood,
that is, child rearing based on rational practices and with scientific support. For
the forward-thinking mother, summer became a time to provide "freedom, sunshine, and increased health."[54]

Foremost among the many challenges facing even the brightest and most
modern of mothers was how best to dress their little ones. Fortunately, the Bureau of Home Economics, a division within the United States Department of
Agriculture, could help. The prescription was the sunsuit. For sickly children,
even less clothing was better—exposing more skin to absorb more light—but
for most, sunsuits could allow playtime to become healthful. One pamphlet
told mothers to modify preexisting attractively patterned clothes. Excessive at-

tention to style, however, was a mistake, lest "the real object of the suit is over-looked. If it does not allow the sun to reach a large part of the body, it is not a sun suit." A 1930 pamphlet offered more advice on the matter, telling parents that ideal outfits easily converted from cooler to warmer conditions and that "suit ensembles are not only convenient and comfortable; they are healthful." In 1929, suggestions from the Bureau of Home Economics reached a broader audience in *Washington Post* articles. Sometimes, according to the newspaper, a "short-sleeved bloomer dress" would work; other times, children needed a sleeveless dress or romper, and at still others, shorts held up with a thin porous material or a set of straps. In all cases, the bureau celebrated the value of play outdoors and a dark tan.[55]

For less enterprising parents, Climax Sun Suit Company had a solution. As the marketing department pointed out, "[There is] something new under the sun, something that will bring new life to your children." Climax's fashion, spe-cially designed for health, would permit youngsters a bit of time with "Doctor Sun." Vanta Baby Garments was similarly enthusiastic about its product, as was Carter's Infant's Wear, which claimed its mix of fabrics had gained the sanc-tion of the Public Health Service, women's magazines, and the New York Ma-ternity Center. With a drawing of an adorably cowlicked youngster, the Hecht Company, a department store, told parents that Snookums, the "inimitable baby star of motion pictures," had followed his physician's counsel with great results: "[His sunsuit is] why he's health[y] wealthy and wise and shows no sign of the strain of his profession."[56]

Though much discussion about clothing focused on the needs of children, sunlight was supposed to be a source of health for people from youth to old age. According to a doctor quoted in the *Chicago Daily Tribune*, theirs was the "dirti-est age in the history of civilization," and it did not help that clothes had severed people from light and air. Travel in "sun-proof boxes" to walled-in rooms had made a race of "sun dodgers"; polluted skies and an "armor of clothes" made all this darkness worse. That was the sort of sentiment that led Coblentz at the Bu-reau of Standards to investigate the ultraviolet permeability of different fabrics. He concluded that lighter colors and thinner weaves were ideal. Palm Beach Suits went even farther, producing outfits for men that it claimed did not block the "vim-giving" rays. According to ads, tests conducted at the Institute of Ap-plied Psychology in Berlin had proven as much.[57]

Innovative, health-conscious weaves could help in the absorption of sun-light, but to many, new designs in clothing held more promise. In 1928, the au-thor of the *Los Angeles Times*'s "Health and Diet" column wrote of the modest bathing costumes of his youth, when boys wore three-quarter sleeves, and women "queer bloomer suits." Fortunately, those days had passed, and the single-piece suit had become the fashion on America's beaches. But gender distinctions in fashion choices remained, to the disadvantage of men. By mid-decade, Leonard Hill, an Englishman, had begun speaking out about the smart clothing that women had started to wear. He thought that lower-cut blouses were good; that artificial silk, permeable to tanning rays, made sense; and that spiderweb stockings helped health. According to articles in the *New York Times*—and one that appeared in the *Chicago Daily Tribune*—Hill believed that "with modern methods of education and constant [sun] exposure women seemed to be becoming the hardier sex."[58]

J. W. Sturmer, dean of the Philadelphia College of Pharmacy and Science, gave an American voice to Hill's argument in an authoritative 1929 lecture later published in the *American Journal of Pharmacy*, and republished in the *Smithsonian Report* for 1930. According to Sturmer, in prehistoric times, primitive men probably went around in nothing more than a modern-day sunsuit. In time, less healthful attire became the rule, but recently, in conjunction with modern designers, women had begun to move past their overdressed days and as a consequence were growing stronger:

> The most transparent textile is a loosely-woven rayon or artificial silk, which explains why the sex known during the Victorian era as the weaker sex is on the way to becoming the robust sex. The girls are getting more solar radiation, hence more vitamin D. . . . The feminine dress of to-day is in far closer harmony with the newer facts pertaining to irradiation as a health measure than is the modern attire of the male. Women have gone far, since Civil War days, in so changing the fashions as to provide for plenty of sun irradiation; and in this span of time the men, alas, have made progress merely to the extent of having shaved off their whiskers and exposing their chins to the sun. Only on the beach at the seashore does the pipe-smoking sex get an even break when it comes to solar radiation. What should be done about this is a matter beyond the scope of this discussion. We give it up.[59]

Sturmer and Hill represent another part of the sunlight discourse that treated matters of gender: women as clothing pioneers. This thinking, part paternalistic, part sexualizing, was summed up in Evans's writing when he quoted Dr. E. Frieberg: "Men are ignorant, opinionated and pigheaded. They will not learn." Women, the argument held, had bucked convention with lighter, looser-fitting clothing that permitted ultraviolet exposure even in winter. In 1926, Chicago's health commissioner provided statistical evidence that women's fashions left them better off: tuberculosis death rates fully 15 people per 100,000 lower. The argument was not aesthetic or moral; it was medical. The commissioner asked, "That doesn't look as if low necks and short skirts are unhealthful, Does it?" His advice beautifully mixed gender with health: "Roll 'em down, girls, wear your dresses thin and sleeveless, with the skirt as short as you please, and fear not. If a policeman doesn't get you there is no need to worry, least of all about the doctor." Authors cast their arguments as part of the battle between fashion and health and protested the "great to-do against the scantiness of women's clothing." Some time prior, prudish tastes would have won, but not in the '20s and '30s, when "the human family ha[d] been awakened to the necessity of preserving its well-being."[60]

By most accounts, the 1920s were a generally liberalizing time in America. In spite of considerable conservative opposition, large groups of citizens, young and old, were behaving in ways that before the Great War had been unthinkable. Freer fashions grew popular in this cultural climate, and legitimacy did not just come from women's magazines and clothing designers, but also from authorities like Sturmer and Evans. They and other health advocates proclaimed that too many moments outside were opportunities wasted. With the aid of sunlight, the body grew healthy, but the silly—and dangerous—commitment to conservative fashions had stopped the process of light absorption before it could start. Clothes, for the harshest of these critics, little more than unnatural trappings of overcivilization, were an enemy to health. Humanity had paid a great price for its alienation from the natural environment.[61]

For one group of Americans, bloomers and stockings were too long, rolled up or down, and no blouse could be cut low enough. Even the journal *Electrical World* acknowledged nudism's appeal in its article for salesmen about sunlamps. According to the serial, "an able physician" from the Metropolitan Life Insurance Company argued the case for discarding clothing: "Man has lived on this earth for many many thousands of years, but he has worn clothes for only a comparatively short time." This choice had repercussions: "It seems that man has suffered somewhat in the change, and perhaps it were wise to make up for the defects by a different manner of living. It would be well for man to stop being a hothouse plant and take his proper place in the sun." This sort of thinking led some far from technology; they wanted the precivilized alternative. The early 1930s witnessed an increase in nudist enthusiasm in America. Though it never reached the heights of popularity that enthusiasts predicted, American nudism is instructive because it was at the outer edge of antimodern sunlight sentiment but well in line with much of what the sun celebrators had to say; nudists bared their skins in protest and for health.[62]

The apparent leap from short hemlines to no hems at all was really not much more than a step. In 1926, the *Washington Post* published an article entitled "The Eternal Question—Clothes." The answer may not have been nudity, but women's desire to shed layers had certainly made them aware of the comforts of less clothing, and scientists, doctors, and physiologists had encouraged everyone to take closer note of ultraviolet radiation. It was only a matter of time before nude sunbathing was recognized for illness prevention. According to Frank McCoy, writing for the *Los Angeles Times*, the future had come by 1931: nudism was recommended for a host of illnesses. With women having blazed a path to new sun-sensitive dress, there could be little doubt that men had to follow suit and unsuit. Clothes, for Leonard Hill, were one of civilization's great evils, so this new and growing group averred, get rid of them.[63]

There was another factor legitimizing nudism: America was not alone in advocating it. Germany, a great repository of scientific and medical knowledge and a leader of the sun cult, had begun to embrace nudism in the 1920s. By the early 1930s, one book estimated a robust movement of four million in Germany, four times the total nudist population of the rest of the globe—with other pockets of enthusiasm in France, the United States, Britain, and Russia. While this claim of widespread German nudity was likely exaggerated, it was not the only suggestion that the nation had become a launching point for a broader movement. A *Washington Post* article referenced *The Way to Strength and Beauty*, a film that an American author claimed "ha[d] filled Germany with a boundless enthusiasm for sunshine and exercise and nudity." More chronicle than narrative, another writer commented, it had "no story to tell, but only a lesson to teach." With Germany's quite public movement, American counterparts could feel they were not alone. While all of these accounts might have been overblown, America's awareness of German nudism was real, and that sense of kinship was what made Jan Gay, author of *On Going Naked*, feel that she did not have to steal away to take it off. A trip to Germany had opened her eyes to semipublic nudity.[64]

Though less a model than Germany, England offered the other international nudist example. The British journal *Sun Bathing Review* suggested that national ignorance contributed to the common belief that nudism was a fringe, amoral movement. In time, that would change because the "modern bathing costume" was priming the nation. The argument concluded that, though patience was hard, English nudists had little choice but to wait; if they attempted to acquire a political association like nudists in Germany or sought "spectacular public- ity" like naturists in America, there would no doubt be a crippling backlash. But two years earlier, the call had been a bit different; it must have been hard to avoid sensationalism or politics with articles that declared, "THE NATION IS AILING! The standard of physique is declining; most of us know that we can and should be healthier and happier. Civilisation and industrialism compel us from childhood to develop our brains to the neglect of our bodies. One of the root troubles is insufficient light and air."[65]

In America, nudist thinking borrowed from *Sun Bathing Review* and took inspiration from its German progenitor. There were two primary authorita- tive new world sources: book-length tracts and the *Nudist*, a journal published under auspices of the International Nudist Conference. The serial had a broad agenda, but much of it revolved around advocacy for a new Magna Carta that would, once and for all, free followers from the overbearing eyes of conserva- tives. Its three planks demanded some park and beach space, free access to the press, and the elimination of legislation banning nudism. This new declaration of rights, true to the seven-hundred-year-old document from which it drew its name, would bring justice to many deserving citizens.[66]

A few authors writing in the *Nudist* counseled acclimation to light and scien- tific tanning. Others maintained that artificial sunlamps could substitute for the real thing when necessary. But far more than regular sun worshippers, nudists were concerned about modern living. Civilization had made free men slaves to "fashion, to style, to inherited ideas, to priceless heirlooms of outworn cus- toms and superstitions." These bondsmen felt trapped by the too-fast tempo of modern life, which pressed sufferers to the breaking point. With minds con- stantly racing, people tried to repress sexual feelings but failed, "tormented with thoughts of sex matters, developing from the romantic and shy to the fiercely li- bidinous." In obscuring the sexual organs, clothing had mystified the body, the root of sexual deviance.[67]

For nudists, nakedness solved all these problems. It broke down sex solidar- ity—which divided humanity—eliminated false modesty, and released men and women from the sexual tensions that usually crushed them. Ilsley Boone, managing editor of the *Nudist* and executive secretary of the International Nud- ist Conference, reminded readers that there was nothing natural about clothes. Indeed, it was not until Adam and Eve transgressed that they reached out and grabbed a fig leaf. The time had come to get past that indiscretion. For some, freedom from clothes permitted a more exciting but more controlled sexuality; for most, it disassociated sex from nakedness, and in doing so, reduced sexual desires entirely. The consequences of the nudist alternative were profound:

> Clothing has built up an entirely false sense of shame, of pseudo-modesty, of secretiveness accompanied unavoidably by erotic curiosity, and worse than all else a concept of nudism that is tantamount to obscenity and suggestive only of sexual satisfaction. No other animal possessing a brain has so completely debased its functioning, as has man.[68]

Returning people to a natural state of affairs, the movement further claimed, would slow the tempo of modern life, "sanit[izing] the mind and impart-[ing] peace and poise." It not only meant freedom from clothes; it meant self-liberation. Once unchained, people would become more open generally. All the many trappings of modern life, including drink, processed foods, and drugs, distracted from good, clean, natural living. In one *Nudist* article, the movement acquired an explicitly emancipatory tone: "The day of physical slavery is past, but psychic slavery is still almost universal, slavery to fear, prejudice, greed, envy and endless customs, conventions and tradition which rob life of its spontaneity and joy."[69]

These enthusiasts, like some of their sun-celebrator kin, rejected modern life, holding that "things invented and manufactured are useless and often harmful." These harms encompassed the trappings of modern life: "Among them are unnecessary and unhealthy clothing, useless structures, built largely for show, ugly and uncomfortable furniture, much trumpery bric-a-brac intended for decoration and many superfluous and injurious kinds of food." Indeed, Boone averred, until recently people had shown a willingness to bare some skin: workers stripped to the waist while plying their trades, and in colonial times, it was not uncommon for people to wear less clothing and walk around naked in their homes. It was only within one hundred years, that "the modern clothing industry, with its metropolitan emporiums of style," made skin taboo.[70]

Humans, Boone and his cohort continued, were natural animals with primitive wants. When permitted to live in accordance with that nature, they became good and peaceful, but civilization had carried man "farther and farther from the normal." Modern culture, the nudists said, "is obliterating all [our] natural instincts and must be regarded as a deviation." Social animals cared for each other; the human, alienated from nature, had lost this communal spirit, becoming a "predatory creature" who used maxims like "survival of the fittest" to justify "rob[bing] his neighbor, and destroy[ing] his own brother." Here, the thinking resonated with Luckiesh's. Humanity had resorted to "calculation, discrimination, and constructive reasoning to great advantage." It made productive machines and new forms of government. Attending those benefits of "so-called civilization," though, was every form of evil that simple communities would never have produced. The result of all these bad choices had been catastrophe; they were the reason for the Great Depression, the "mad debacle as the most civilized nations have enjoyed." In *The Nude Deal*, a self-described "Jolly Gesture 'Gainst Bulky Vesture" which joined bad, folksy poetry to a critique of capitalism, Art Eastman also saw a clear socioeconomic benefit to clothes-lessness. Nudism was the "great leveler": "In Bali or Boston a man or woman without clothes is without class distinction. One may join the navy and see the world or join a nudist colony, and see his neighbors as they really are."[71]

Nudism, as a philosophy, did not offer a compelling explanation of how nakedness would change people. Rather, it sold a new mode of living, rejecting modern as artificial—and therefore corrupt—and hoping followers would turn themselves over to a more natural way of life. Freudianism, a postwar sensation in America, helped shape this nudist challenge. While most who turned to the Austrian for guidance in matters of the mind misunderstood much of what he had to say, they did try to get it right. Repression, according to these American Freudians, crippled. When unable to find natural, normal expression, primitive wants became dangerous tendencies. Nudists argued similarly. According to one account, the repression of "prurient curiosity" by an overbearing society was, undoubtedly, "one of the causes of the neurotic and hectic character of our civilisation." In "Balance," the *Nudist* editorialized that overbearing "ancient moralists" had caused mental instability, unsettling healthy minds and yielding sexual deviance.[72]

In its celebration of natural needs, nudism resonated meaningfully with sun worship. Clothes mirrored darkness as a great denaturalizing menace. There were other ways, however, in which the two movements fused even more fully and seamlessly, with nudism relying on the substance of sunshine thinking in advancing its argument. The covers of early *Nudist* issues contained the descriptive, "Health building through sunbathing, diet and exercise," and before long, the magazine's publishers turned to a less controversial title, *Sunshine and Health*. A number of the early organizations advocating the bare lifestyle called themselves sun or sunshine leagues, and hoping to avoid cultish connotations,

I. O. Evans's book *Sensible Sun-bathing* decided on the euphemism "complete sunbathing." This movement, he claimed, had thrust mixed-gender, nude tanning to the forefront of public attention.[73]

But the relationship between the two movements was not just rhetorical; nudists believed that sunshine was a part of that natural life they sought to recapture. Proponents Charles and Frances Merrill sent surveys to psychiatrists, psychologists, and doctors asking whether they sanctioned the nudist lifestyle. For all groups, the response was mixed. What baffled the Merrills most of all was the failure of health experts to fully embrace the role of nudism in making healthy bodies:

> The lay public in America, being addicted to sunbaths and having fully accepted the gospel of ultraviolet rays and vitamins—even though not always clear as to just what those rays are, or how they perform their miracles—is apt to take it for granted that medical science and health authorities in this country will likewise welcome the cause of nakedness, though perhaps with reservations regarding unsegregated nudity.

According to the authors, conservative doctrine and the fear of a judgmental public had closed medical eyes to good reason and sound science. True "balance" had as much to do with the physical as the mental. Much as overbearing moralizing damaged minds, alienation from the physical environment crippled bodies. The nudist story was apocryphal, telling of a golden age when freedom was king: "For millions of years man accepted life as it was," eating whatever struck his fancy, bathing in light for health, and sleeping wherever sleep came.

> Man's whole person was tanned with the sunlight and washed in the rain. In the millions of years of man's development he did not even cut his hair or wrench out his beard. He accepted nature as she made him. In the long process of development many obscure balances were built up. The strength of the human being's bones depended partly on his diet and partly on the amount of sunlight that fell on his skin.

Here nudists made their outdoor philosophy the complete fulfillment of suncelebrants' vision.[74]

Other articles in the *Nudist* played off similar themes. Upon returning from the "filthy, blood-drenched battlefields of France," Harry Ellington Curtis set out to restore his health and found modern medicine little help. Fortunately,

Mother Nature came to the rescue. The shattered veteran headed to Mexico, where he learned lessons from the Tarahumare Indians, a people he mythologized as being among the healthiest and longest-lived races, with more than a few centenarians in their number. The Tarahumare could chase down rabbits and deer with ease and run two hundred miles or carry a hundred pounds while traveling great distances.[75] The lessons Curtis learned made him a sun worshipper and convinced him that man had a lot to learn from animals when it came to health. In another article, reminiscent of the tale of Rollier's dog and the animals at the London zoo, Boone told the story of a little slum dweller who, in spite of her dark and dismal surroundings, had captured top prize at a London flower contest. She had managed to find the few rays of sun in her gloomy home and moved her plant to follow that light throughout the day. Sadly, the girl did not have such a mindful caretaker, and "the little one did not know that the sunshine was needed even more to bring the flower of her own youth to its most perfect blooming."[76]

It is hard to gauge the full scale of nudist enthusiasm. In 1934, one author put the total American circulation of the *Nudist*, available on the newsstands of 540-plus cities, at 100,000 per month. Perhaps more notable and probably more reliable, the serial repeatedly added to its tally of sanctioned nudist leagues, a list impressive for its rapid growth, absolute numbers, and geographic distribution. In the middle of 1933, the second year of the *Nudist*'s run, it reported nine leagues or groups and four farms or camps. They were all in the East or Midwest except for one in Sugar Land, Texas, but besides that, the locations were more notable for their differences than their similarities, ranging from Florida to Massachusetts and varying from metropolitan New York to Kalamazoo, Michigan. At the end of the year, the directory had grown from thirteen to forty-four and all geographic boundaries had eroded: California had taken its place as a hotbed for nudist enthusiasm. By January of 1935, the number reached fifty-nine.[77]

Predictably, not everyone was happy about this proliferation, and the movement consistently had to defend itself from conservative attacks. In "Signs Along the Way," *Nudist* author Donald Thistle spoke hopefully of the future. Until recently, he said, Chicago had been very strict about its modesty laws, but by 1933, dress restrictions were going unenforced: "Day after day you may see bronzed, athletic, beautiful physiques on Chicago's beaches, physiques which are exposed to the soothing, healing influence of sun and air and water." More exciting, there was a legislative push to legalize nude bathing, which presaged a new future for "Puritanic America." Articles in the *Chicago Daily Tribune*, the *Washington Post*, and the *New York Times*, reported on a recently proposed ordinance to set up a sex-segregated stockade on Lake Michigan for nude sun-bathing. Advocacy for the measure in 1932 had put Mayor Anton J. Cermak in a tough political position; he saw the good in sunbathing, which he said, made happier and healthier citizens, and counted himself a practitioner, but he was not a nudist. The mayor, according to the Chicago paper, was against the law but unwilling to oppose it head on. He had asked Alderman John P. Wilson, chair-man of the city's committee on playgrounds, to undermine a resolution on the pen, because "nude sunbathing is not done in this country, and the city never intended that it should be started in Chicago."[78]

The mayor was being a little disingenuous with his pronouncement; the jury was still out. With nudist leagues proliferating, consternation grew and courts across the nation had to decide matters of clotheslessness. Often, they ruled that nudism practiced on private property could not or should not be prohibited. Unwilling to let the courts decide based on existing legislation, former New York governor Alfred Smith and the Legion of Decency sought to criminal-ize nudism and ban movies sympathetic to the cause. In San Francisco, the ap-proach to regulation was far different. According to the Los Angeles Times, with a 400 percent rise in nudism in a month, the health commissioner of the City by the Bay felt he had little choice but to accept the arrival of a nudist camp and had taken necessary steps to regulate it. Rules expressed concerns over decency and required walled facilities, sex-segregated bathing, and board-of-health over-sight of hygienic provisions. There was even something in the regulations for the wives of health inspectors, who could refuse to let their husbands supervise the nudist refuges. Though neither the Times nor the San Francisco Chronicle hazarded a guess about how many visitors the area had—and 400 percent of a tiny number is still small—the nudists were capturing the public's attention. They may have been oddities, but they were oddities discussed at the highest levels of municipal government, oddities politicians now thought worth regu-lating.[79]

At Chicago's Century of Progress World's Fair, the Zoro nudist colony caused quite a stir. Zorine, daughter of the sun, was the wife in a staged naked wedding at the behest of a publicity-hungry concessionaire; within a week, the faux hus-band and wife were facing charges for their stunt. Later, the Zoro colony was invited to give a clothed show at the Spanish Village, but within days, mem-bers were again in trouble for public indecency. Zoro's legal troubles were gone by the 1935–36 California Pacific International Exposition in San Diego. The colony had become a fully sanctioned exhibit. In fact, the nudists had to cover up with bras and g-strings (it is unclear if they were similarly clothed in Chicago), but in one of the fair's promotional postcard collections, that information was nowhere to be found. Instead, it boldly featured the nudist "amusement." The ongoing show included everyday activities, from eating to athletics. One short movie showed a ceremony performed by the queen, in which she arrived in a robe, was decloaked by her attendants and presided over the assembled. This time, the *Chicago Daily Tribune* report was more favorable. From two thousand miles away, it noted the exhibit's success: 125,000 visitors as of September 8, 1935—only the exposition's midget colony was a bigger draw. Though protests surely continued, nudism had gained a measure of acceptance.[80]

The primitivist elements of the Zoro exhibitions were evident, from marriages in foregone, make-believe worlds to postcards depicting cherubic women and aged men in idealized settings.[81] Its nominal relationship to Zoroastrianism, the ancient Persian religion known in part for its sun worship, reinforced "the sun daughter's" association with the past. Certainly, there was little about these shows that accurately represented humanity's true state of nature and probably even less that accurately depicted Zoroaster's teachings, but that was not the point. Nudists like Zorine connected themselves to invented times when humans lived healthy, natural lives in peaceful harmony with their environment. This mythic sunsoaked past was an ideal foil for a corrupt, dark age.

It would be a mistake to suggest that a fourfold increase in San Francisco nudism or a circulation of 100,000 for the *Nudist* meant that this movement was sweeping the nation. But although nudism may have been exceptional, its limited ranks do not mean that it was hidden from public view or that it found no place in American culture. The thinking behind it resonated with much of what many mainstream Americans had to say about cities, clothes, and civilization. Moreover, while people did remain dressed, less complete changes to American attire in this time period were enduring and widespread.

Publisher Bernarr Macfadden was an early twentieth-century health enthusiast. His personal origin myth was a common one: a sickly childhood followed by a startling revelation about how to be well. He loved pumping iron, hated white bread, and liked sex—a lot. Macfadden was no outsider. For a short time, Eleanor Roosevelt was the editor of one of his magazines on child rearing. He stayed at the White House during the FDR administration, met with Mussolini, and harbored a not outlandish dream of becoming the Republican nominee for president. At Macfadden's peak, his publications, from daily newspapers to monthly magazines, had a circulation larger than Hearst's. In short, Macfadden was a health enthusiast with a following.[82]

His greatest publishing triumph was *True Stories*, which contained readers' tales—often tall—but throughout his work, Macfadden saw his mission as that of public educator. Between the turn of the century, when he began publishing his first magazine, *Physical Culture*, and the 1930s, when his media empire reached its full form, he developed a personal theory on the importance of natural living. His advice peppered his daily newspaper, the *Graphic*, and other publications, but nowhere was it more explicitly spelled out than in *Physical Culture*, which peaked in 1933 with a readership over 300,000.[83]

Physical Culture spoke often about the merits of sunlight and echoed the warnings of health experts. It discussed animals, the Greeks, and the dark ages. It talked about rickets and tuberculosis. It called sunlight the second most important natural force, next to oxygen. One article title extolled the "New Super-Food for Vitality," vitamin D. There were plans for sunshine playgrounds, and arguments that light would bring a long life. Always looking to smash conservatism, Macfadden even wrote about how modesty did little good for principled lives in editorials like "Clothes and Morals" and cast his lot with "complete sun-bathers" in "Nude Cults Are Smashing Old-Time Prudery."[84]

Macfadden's love of sunlight indicates how suddenly the movement caught on and how widely it spread in the 1920s and 1930s. His *Encyclopedia of Physical Culture* went through multiple editions. Its first in 1912 celebrated fresh air and made passing reference to sunlight, but as little more than adjutant, a friendly helper that might stimulate the nerve centers or help the body absorb outdoor vitality. Sunbaths received no mention in the treatment of rickets, and tuberculosis care focused almost entirely on fresh air and exercises. By the later editions of Macfadden's encyclopedia, the highly circumscribed role for sunlight was gone. The eight-volume compendium contained extended articles entitled "Sunlight a Foe to Disease" and "Sun Treatment by Artificial Means" and offered a laundry list of conditions ultraviolet light might cure. He pressed his case with references to nude sunbathing and exhortations to pay close attention to sunlight obstacles like smoky skies. Though Macfadden had long feuded with the medical establishment, he used its language—if not its dogma—in talking about the benefits of light for patients suffering from infections, lowered hemoglobin levels, or lymphatic problems.[85]

Macfadden was popular, but I do not mean to suggest that he could single-handedly shape tastes; in fact, his magazine might not have changed many minds. When the publisher pushed his natural-living case too hard in the 1940s at a time when the public was less committed to physical culture than to modern medicine and was in any event most concerned about a world at war, Macfadden lost his empire. He was so popular in the 1930s because he was so in touch with his day, not because he convinced the world he was right. Sunlight enthusiasm had spread far.

By the mid-1930s, the movement had reached its full form. The charges were bold and clear: natural living was good, and modern living could be dangerously dark, overdressed, and artificial. To overcome the challenges endemic to urban life, people needed better housing and smarter parenting. Though San Francisco health authorities might have thought the nudist lifestyle was unclean, a great many Americans thought winter wear was much less hygienic. A coat of tan, medical science counseled, was key to health. And never before had humanity more needed health—new threats to the race required better conditions for improved breeding. By the mid-1930s, the problem was clear, and solutions were of the utmost importance. Nudism was one, a self-avowedly antimodern critique.

It was unlikely that "New Gymnosophy" advocate Maurice Parmelee would see the future of which he dreamed:

> As regards the ultimate course of social evolution, one of the most important results from the universal adoption of gymnosophy would doubtless be to encourage a tendency to move away from the cold regions into the temperate and tropical zones. This would be in strict accordance with man's biological nature, because he is an animal adapted to a warm climate. No mammal without a thick fur is well adapted to a cold climate, and man with his entirely bare skin is the least adapted of all.[86]

In the end, however, Parmelee did get some of what he wanted. The north did not empty out, but the tropical south, with its beautiful sunshine, was coming alive. Many decided to relocate; far more embraced a half solution: sunshine vacations. For the rest of the year and the rest of America, who could not get away at all, there was a new development, far from the nudist vision: fortified foods that gave nature's promise. Both improved diet and an afternoon at the beach, experts counseled, had benefits, but extended time outside was beyond the reach of many families. Sunshine had become an expensive commodity, just the sort of inequity that tenement reformers had argued against; but a glass of milk, which now contained the minimum nutrients for health, was a remedy that could defuse the arguments of those who demanded some time in the sun for every American.

5

Climate Tourism and Its Alternative

The *Awful Truth*, a 1937 Cary Grant comedy nominated for six Academy Awards, tells the story of Jerry and Lucy's difficult attempt at divorce. Uncertain of their feelings for each other but in a marriage wracked by suspicions of infidelity, they look to move on. In the film's first scene, the clock strikes eight at the Gotham Club, and Jerry calls upon Hank, one of the club's attendants, to get the sunlamp ready. The protagonist, whose wife thinks he has been off in Florida for two weeks, is pale, clear evidence that he has lied about the trip. Hank advises an initial treatment of a quarter of an hour but is rebuffed: "Aaaaa, fifteen minutes nothing, I gotta get a deep Florida tan if it takes all afternoon. Give it a gun. All aboard for Miami, Palm Beach, and points South." In the next scene, Cary Grant is deeply—and unambiguously, though the film is shot in black and white—brown. Lucy, who has been with her voice teacher, a man Jerry does not trust, enters, glowingly pale and happy to see him after their fortnight separation. She proclaims, "Darling, oh, how grand and brown you are."[1]

Cary Grant's character knows that a trip to Florida means

a coat of tan. His wife does too. In the mid-1930s, Floridians might have acknowledged that Stuart Chase's fellow sun worshippers were wise and the visiting nurses who counseled mothers important, but residents of the Sunshine State were aggressively asserting that there was nothing quite like their exceptional climate. Californians, too, said their state was "unique," as did Coloradoans, New Mexicans, Arizonans, and Texans. In fact, America sparkled with uniquely sunny locales. Some celebrated that the sun was a truly democratic healer, shining anywhere and on anyone, but even the most ardent nudist had to admit that not all climates were ideal. That being the case, those places with the best light could hold themselves up as new destinations for health seekers. If the sun could not come to Minneapolis, Florida's municipal boosters proclaimed, Minnesotans should come to them.

Floridians claimed that their state was close and affordable to nearly all of America, but in reality no vacation destination could fully meet American needs. People worked. People were poor. People could not travel. The eventual solution to sunlight woes followed two distinct paths: for those who could, travel; for everyone else, four glasses of vitamin D milk daily. Low priced and widely accessible, dairy fortified with the important health element promised to bring sunlit wellness to anyone in need. By 1935, vitamin D fortification was becoming commonplace, a shift that would forever reshape sunshine concerns.

One reading of this story could be that tenement reformers, who for decades had sought a little light for everyone, had won: a democratic solution, by which everyone could get well, had bested commercialism. That understanding would be a mistake. Business had found a way to produce a cheap sunshine substitute, but that did not fix many problems. People were supposed to get away and outside. Early sunshine advocates and later health professionals counseled a return to more natural conditions, and that could not come in a bottle.

By the 1910s, southern and western destinations had begun to realize that the climate tourism that drew ailing easterners in search of health was becoming more of a liability than an asset. Linking their fortunes to the ill—often terminal tuberculosis patients—had its perils, and as the popularity of places like Southern California began to take off, they looked to sever their tie with the great white plague. They did not, however, shed their association with a salubrious climate; "Sunny California" would stay. Across the country, locations sought a similar association with little concern for the plausibility of those claims.[2]

Sunshine and Grief in Southern California, a short book published out of Detroit by an unnamed "Old Promoter Forty Years in the Field of Real Estate," predicted bad things for California. The state's prodigious rise would soon end, undone by its lack of substantial selling points and sole reliance on a lovely climate to generate enthusiasm. Its cities were poorly constructed, its lifestyle idle, and its people mentally unstable—prone to suicide and drug addiction. Eastern transplants would grow tired and would stop coming when "riches and leisure, idleness and pleasure have played their part on this sun-lighted stage and the call comes for a change of scene and a season of substantial activity in a more progressive community." For those interested in a sunny lifestyle, bright southern and eastern alternatives would see opportunity and promote themselves as substitutes.[3]

In songs like "California Sunshine," "Golden California Sunshine," "Make Up Your Mind to Wind Up in Sunny California," "Sunny California: You're Home Sweet Home to Me," and "The Sun Shines Bright in California," lyricists countered the pessimism of *Sunshine and Grief*. The state's climate was an indisputable part of its irrepressible beauty, and that was a part of its appeal. In "Oh! California Sunshine!" the argument was more developed:

> Oh California Sunshine,
> How Wondrous are thy rays
> There's magic in thy glimmerings
> There's healing in thy touch
> Thy light o'er all enchantment brings,
> There's glamour in thy golden days

Florida sunshine received its own lyrical panegyrics in "Sunny Florida" and "Tourist's Song." In this tropical paradise, it was St. Petersburg, "The Sunshine City," that marketed itself most aggressively. On the cover of the sheet music for one of the songs dedicated to this city of such great natural bounty, publisher Mrs. Leroy P. Naylor called it "the best place under the sun."[4]

The "Old Promoter" was no more right about Florida, which he claimed had been a land of opportunity in the 1920s but was fast becoming an irredeemable bust, than he had been about California:

> When the curtain was finally drawn on the scene of the wild orgy of speculation and the sadder but wiser men and women returned to their old homes in distant states, the irresponsible land boomers, subdivision men and high pres-

sure salesmen sought other sections of the country where they might continue to pursue soap bubble schemes.

Predictably, in its book *Florida: Empire of the Sun*, the state's hotel commission did not agree. The commission did not see the boom, which had increased the state's population to 1.5 million, ending any time soon. Florida had become an empire, not just because climatic advantages were helping it to predominance in industry and agriculture but also because it was the Sunshine State, "where the beneficent and healthful effects of year-round sunshine are so justly famous." As evidence of this wondrous environment, the commission enthusiastically reported of the *St. Petersburg Independent*'s promise to give the paper away on the rare day when the sun did not shine. In nineteen years, publisher Major Lew Brown had eaten the cost of his daily only ninety-seven times. With that record and the healthfulness it presaged, it was no wonder, according to the boostering book, that history could find no parallel to the state's sunlit rise: the California gold rush and southwestern oil boom were little match.[5]

Though there had been some enthusiastic talk of sunshine tourism prior to the mid-1920s, it was not until then that the real dollars started to come. Medical evidence had supplied the necessary proof that a trip to a sunny climate was an unmitigated good, and newspapers spread the word. One *New York Times* article asserted that, on the surface, winter trips west and south might seem like a "social migration," "a custom of civilized man." But really, this restless movement came from within, the result of "an actinic hunger," an instinct "as the mechanistic heliotrope turning its face toward the sun."[6]

Of course, cities other than St. Petersburg often played up the evidence that they would build health; Miami boasted uniquely powerful sunshine. Confidence in his product led the city's mayor, E. G. Sewell, to extend invitations to Chicago's shut-ins. A 1934 *Tribune* article claimed that Ponce de León might not have found the fountain of youth but modern Americans had discovered its secret at the beach. Readers were told of the Florida politician's recent visit to open a travel information office on North Michigan Avenue. According to a subsequent issue, "America's winter playground to the icebound northerner" was a wonder, which was why Mayor Sewell

might be persuaded to admit that Florida sun has done more for public health than have all the clinics, hospitals, and research foundations in existence, and to advance the thought that if all America took advantage of what Miami has to offer in the winter time virtually all sickness and disease would perish from the western hemisphere.

If only everyone knew. In 1931, Miami began to sell itself more aggressively and expected 100,000 tourists to take advantage of low-cost summer vacations that would confer considerable benefits. By 1935, the focus was on winter tourism, and an estimate in the *Tribune* (no doubt inflated, but nevertheless instructive about attitudes toward southern destinations) held that if advertising could tell the full story, the city would be awash in 5–6 million visitors per year.[7]

Around 1925, Coral Gables began an aggressive campaign to tell its part of the story. This was not a soft sell; close to Miami, the city offered access to culture and sophistication, but far better, it offered a healthful lifestyle in a beautiful setting within reach of nearly every American, forty-eight hours or less from seven-eighths of the nation. Promotions by the Ayer Advertising Agency mixed common boosterism, financial opportunism, and medical enthusiasm: "Coral Gables . . . Miami . . . Florida—words synonymous with health . . . and wealth!" No short-term wonder, this was a suburban ideal whose growth would not flag because it had superior facilities and a smart plan. It was a dream city, "a logical center for southern Florida's great industrial expansion," with something for everyone: "Bronzed, erect old men. Women delighting in new cream-and-rose complexions. Round and brown children. Handsome full-figured youngsters." They were all evidence of the "extraordinary vitality and superb health that come from living under the tropical skies of Coral Gables." In advertisement after advertisement, "golden sunlight," "everlasting sunshine," and "that ancient life-giver the sun" were presented as nature's gifts to visitors and sure moneymakers for investors.[8]

A vacation spot rather than a planned community, the Sun-Ray Park Hotel Spa and Health Resort boasted that it was the original Miami solarium. One of the facility's promotional pamphlets talked about a robust outdoor program and cited more sunshine as one of the twelve reasons that it was best. Located at the center of a city becoming a shrine for recreation seekers, it was a modern facility that followed the expert advice of Greek and Roman solaria enthusiasts, an ideal place for the treatment of conditions ranging from arthritis to colds and high blood pressure. Most of all, it was good for its "happy combination of rest, recreation and pleasure, and a definite plan for health building."[9]

Throughout Florida, the refrain was similar. Ads published in northern papers beckoned weary visitors down south to recharge: "Let Dr. Sun Protect Your

Health at West Palm Beach," "Outwitting Winter in the Cities of the Sun," "Give Them the Gift of Sun-Tanned Health!" It was always June; there was endless vitalizing daylight; the tropical sun would bring health to children. One advertisement for St. Petersburg in *Hygeia* reminded readers that these were not simply empty proclamations. Doctors were counseling a trip to the Sunshine City: "Mother Nature quietly does her good work. Many are the records of regained health, of prolonged life, of childhood given the birthright of health and stamina, through St. Petersburg Sunshine!" The town even boasted municipal and private solaria for modern sunbathers. For those interested, there was a free booklet.[10]

One hundred thousand visitors to Miami might have seemed like an impressive number, and 110,000 residents was notable for a city that had not been one of the nation's hundred largest only ten years earlier, but the rise of Southern California in the early 1920s was something else entirely. In the 1920s, 1.5 million people migrated there. Los Angeles began the decade with 576,673 people and ended it with 1,470,516, making it the nation's fifth largest city. Construction followed population trends, and in the peek year of 1923, the city issued more than 62,000 building permits. In the early 1920s, the California All-Year Club came together to try to boost tourism. Within ten years, it led Southern California's attempt to go from summer destination to twelve-month haven. The year 1932 was a tough one for tourism nationwide, but the All-Year Club boasted that its 1,036,730 out-of-state visitors indicated that the Depression had not done the region too much harm.[11]

Of course, not all folks came for the sunshine alone. By the 1930s, California offered countless opportunities. People moved to make money in movies, agriculture, and industry, and visitors most frequently came because they now had family to see; in fact, a survey of visitors published by the All-Year Club found that by far the most common single reason for visiting was family (29 percent). Only 3 percent of visitors said they came for health, and 11 percent for climate. But 11 percent of a million is still a lot of visitors, and many of those seeking a reunion probably saw climate as an alluring subordinate reason for heading west.[12]

By 1932, *Los Angeles Times* authors were talking about the blossoming tourist trade and the common perception that winter in the "Sunshine City"—St. Petersburg did not have a monopoly on the nickname—was an easy alternative to a miserable northern January of blizzards and earmuffs. One article claimed, "From the standpoint of health and recreation, sunshine is the city's outstanding weather element." Edgar Lloyd Hampton characterized the city's good weather as Janus-faced, both gift and curse: tourists had flooded the state, which made it difficult for locals to enjoy. There were two primary reasons for this westward tide, scenery and sunshine, things even a grouch had to love.[13]

The *Times*, as a publication, did not share its columnist's ambivalence. In the late 1920s, it ran a series of ads for its winter supplement, which proudly boasted the sunshine benefits of the western paradise. A six-part magazine series would help inform readers about the ways that ultraviolet light was building a "new and healthier race." This was the single greatest discovery of the time, the serial claimed, and Southern California offered a unique, "powerful curative." The advertisements encouraged readers to order extra copies so they could "send California's sunshine to [their] friends back East!" in order to inform, or gloat, about "the tingling exhilaration of outdoor life in the Southland—its wonderlands of scenic splendor—the hum and laughter of busy, ambitious, happy people at work and play." The serial itself was not quite so overwhelming in its sunlight enthusiasm. As was the case with other annuals, this edition was more a celebration of industry and agriculture than a health manifesto, but it did point to tourism as one of Southern California's top ten revenue generators, and it featured an article by Hampton about a new, healthier race grown in California sunlight.[14]

Contrary to the contentions of boosters, neither Los Angeles nor Florida could claim a monopoly on sunshine; in fact, the City of Angels had competitors within its own state. San Francisco, the City by the Bay, more noted for rolling fog than sunny play, advertised winter scenes dominated by summer sunlight. With cutouts of people in front of beautiful, bright backdrops, it encouraged readers of *Good Housekeeping* to insert themselves into the healthful scenes. It was California's desert, however, that held the state's best claim to ceaseless sunlight. Articles in the *Los Angeles Times* treated Palm Springs and Death Valley as they would sanitaria, places with exceptional atmospheric advantages, which showered visitors with ultraviolet rays, "the genii of modern medicine."[15]

In 1922, Tucson's civic leaders, residents of a relatively small city tucked away in an underpopulated state, established an organization to increase visitors and encourage transplants. In *Tucson Magazine*, the Sunshine Climate Club celebrated "America's Winter Playground" in the "Sunshine Center of America." The club reproduced maps of the desert Southwest, with Tucson, where the sun shined bright 84 percent of the daylight hours, at its center. According to the manager of the club, it had raised $200,000, a great investment because it resulted in contacts with 76,476 people and almost 7,000 visitors, who spent 9 million dollars. Moreover, Tucson was in the midst of a population boom (between 1910 and 1920, its population increased by 7,000, between 1922 and 1929, by 22,000).[16]

Tucson got some of its best advertising from Henry Wright Bell, who sang the praises of his sunshine haven. The *American Magazine* published his "Why I Did Not Die" in 1924. The article was serialized in other periodicals and re-printed as a booklet by the Sunshine Climate Club. Wright's story echoed those of other sunlight enthusiasts. Sick with pneumonia and unable to regain his health, he headed to Southern California, where the climate made him well. Un-fortunately, the improvement did not last. A car accident unsettled his precari-ous health, and bad got worse when a chest X-ray revealed tuberculosis. Though he was concerned about his ability to endure the trip, Wright again moved, this time to Tucson, because he "knew that Arizona has more sunshine than any other accessible spot on the continent." Hard luck brought an uncommonly rainy season, but his doctor reassured him that he had made the right choice: it was the quality of sunshine that determined Arizona's curative power. No light was of better quality than Tucson's, and before long it would again shine. Fortu-nately, Wright said, the doctor was right.[17]

The rest of the nation chimed in as well. San Antonio too claimed that the Southwest was the Far West's equal when it came to sunshine. Ads boasted that the Texas sky offered 168.4 hours of sunshine in January, 166.9 in February, and 210.4 in March—days filled with adventure, air as fine as "old wine," and a blue Texas sky from which "healing sunlight pours over the countryside." Col-orado's not-quite-desert climate could add yet one more layer to its pitch. Not only was the mountain sky clear, it contained a most "benevolent sun," unparal-leled at sea level; put simply, the state was a "land of sunshine and vitamins." At base, these places argued that the nation could be divided into sunshine-debtor and sunshine-donor locations, but self-purported sunshine donors were not be-holden to the truth. With the potential for sunshine anywhere and darkness everywhere a concern, there was no place without a claim, dubious though it may have been. If San Francisco was a stretch as a light lender, western Michi-gan was a leap. Nevertheless, in ads, the "playground of a nation" sold itself as a place to escape "the dust, heat and roar of the city" where the "ultra-violet rays—so beneficial in renewing health and energy—exert their full effect in this pure air." Parents were supposed to bring youngsters for "ruddy cheeks and robust bodies."[18]

The sun even shined on the way to sunshine. In their seabound vacations to the Caribbean, French Line promised patrons a trip that combined relax-ing fun with rejuvenating health. Sunshine "was the order of the day" on modern ships, which offered uncovered sun decks, where it was only "you, the sea, and vitamin D." Another advertisement produced for *Harper's Bazaar*, *House and Garden*, and *Fortune* in 1933 waxed more lyrical: "With games and ancient mysteries, far-darting Apollo is adored. Votaries wrapped in steamer rugs lift up their faces to the Great God Tan." In 1928, the Burlington Route Trains, which traveled from Chicago to the Twin Cities and Nebraska, claimed that, in the company's neverending search for luxury and sophistication, it had found a way to bring passengers ultraviolet light on their trips in a sunporch car, specially glazed for health.[19]

Many secondary businesses grew rich from this new celebration of sunny locales. Sunsuits were of course one, but so was suntan lotion. Unlike later products, the goal was not to keep the skin safe from all ultraviolet light but to moderate exposure. Sunscreens permitted a healthful, toning tan while preventing deep and dangerous burns. Unguentine, marketed aggressively in the late 1920s, reminded prospective consumers, "Sunlight is the greatest of natural tonics," but encouraged them to exercise caution in their exposure: "Be on your guard against cruel sunburn—it strikes treacherously without warning." It was a hazard that killed cells and released toxins: "The right amount of sun is the best thing in the world for you—too much may be very dangerous."[20]

More curious were the products that sought to capitalize on the rise of sunshine in less direct ways. In the early 1930s, Lucky Strike cigarettes ran a series of ads, "Sunshine Mellows, Heat Purifies." Much of the text was repeated, and the picture varied only slightly: it was always a young man or woman sitting on the beach in a bathing suit. These were not simple attempts to sell with sex—though that was, perhaps, a subtext. Lucky Strike encouraged readers to consider their product a tempered tobacco, devoid of the harshness of other cigarettes because ultraviolet-ray toasting stripped their smoke of its harsh bite. Doctors, the ad reminded readers, told people to "keep out of doors, in the open air, breathe deeply; take plenty of exercise in the mellow sunshine, and have a periodic check-up on the health of your body." Other ads told of notable businessmen who called the ultraviolet toasting of tobaccos a great example of the successful marriage of science to business and of the 20,679 doctors who found Luckies less irritating: "Everyone knows ... sunshine mellows—that's why TOASTING includes the use of the Ultra Violet Ray." In essence, much as the sun was a tonic for the body, it was a tonic for Lucky Strikes.[21]

Throughout much of this thinking, sunlight's role remained constant, whether for people or tobacco: it soothed. And, many contended, this high-pressure time was brimming with overworked, shut-in, unhappy laborers who needed soothing. Pale was beautiful in the nineteenth century largely because it con-noted freedom from work, a luxurious lifestyle. When tan became beautiful, the class element remained, but it was not so straightforward.

With work the standard in America and most work indoors, pale no longer meant time free from fieldwork. Leisure meant time outdoors. In its pamphlet *What Everyone Should Know about Ultra-Violet Rays*, Hanovia related that cult-ish sun enthusiasts eager for health had increased in number tenfold within a decade. Each summer sun worshippers headed off to the beach and out to parks, enduring overcrowding for a little time in the "warming, energizing rays of the sun." In winter, needs were similar, but sunshine up north was worth little. That was why "followers of the sun" sailed to the West Indies and Bermuda or took a train down south. Others who had the time and resources headed to Florida, California, or the Riviera—to fashionable resorts, resorts that had "become fashionable only *because fashionable people frequent[ed] them*." Hanovia told its consumers—most of whom were probably well aware of it—that there were some "whose obligations make it impossible to get away, even for a while, to bask in the health building rays of sunlight." The lamp maker was selling a tan for that unfortunate majority who sought healthy and fashionable brown skin but who lacked the ability to get it: "Everyone can now safeguard health with the vital element which sunshine provides to build up resistance, to create the vigor and energy of complete well being!" In its ad "Where Did You Get That Tan . . . ?" Vita Glass featured a businessman who mistakenly thought his ruddy-cheeked colleague had been off in Florida, when in fact he had just been tanning behind Vita Glass at home and in the office. That was a theme in lamp ads, too: Florida sunshine made to order. Part of the goal was certainly health, but the ruddy tan had come to connote sophistication and freedom from work.[22]

It would be an oversimplification, however, to simply morph the pale loaf-ers of the nineteenth century into tanned outdoorsmen of the twentieth. On shop floors lit by type S-2 bulbs and in mine sunrooms, the goal was a more healthy and productive workforce. When he proposed his "practical window treatment" in *JAMA*, A. H. Pfund noted that its low cost was invaluable, as the poor were "the ones most in need of heliotherapy." Elsewhere, programs aimed at improving the public's health tended to the sunlight needs of the working class and the very poor. The Pennsylvania sanitarium system focused attention on the indigent, and camps, fresh air work, and settlement houses catered to those of limited means.[23]

In reality, the goal was rarely complete relaxation or a life free from work.[24] Miners toiled away all year, and the characters in Vita Glass ads may not have reported to factories, but they were clearly workers. The summer vacation was part of a fuller life aimed at greater productivity. By taking time off every year, employees could return to their jobs with greater vigor. According to social critics, the problem with office work was that it was too fast-paced. People could not sustain themselves all year without a rest: "The man who thinks he can work forever without a vacation is a stunt flier in an endurance test." Trips were times to recharge, to bask in the sun, and to regain a natural balance and improved health. In the ad "A Word to Wives Whose Husbands Are Worried about Business," responsible women encouraged their men to take vacations, persuaded of the truth that was being sold them: "His health is his greatest asset, worth any price." While infidelity may have kept Carey Grant from Florida and led him to use his health club's tanning facilities, for others, vitality was one of such places' primary draws. Crystal Health Clubs sold vigor to executives with personalized training, squash, and the "magic of ultra-violet sun rays." The need was profound because "many fail to see impending physical disaster through the fog of business demands." It was important not to wait for treatment: "Before you know it, you're flat on your back . . . an enforced vacation that's expensive and inconvenient."[25]

The association of sun with vigor was settled, and as long as that was the case, tans were good, and beaches were places to go for a vacation. But a move west or south and a life of leisure were not within the reach of most Americans. The vacation was, by its very nature, an exceptional time, a time away. Fortunately, there was another type of sunshine solution. Scientists had identified vitamin D as the greatest health benefit that the sun provided; there may have been others, but they were less understood. Around the time that ultraviolet glass and rickets were discovered, scientists made one more critical finding.

Though they had no scientific theory to explain why, it was obvious to researchers that the Inuit were well ahead of the health curve. According to articles and advertisements from the late 1920s, scientists had solved a great medical mystery when they figured out why Eskimos did not get rickets even though they endured the darkest and longest nights and wore the thickest clothes. It turned out that they ate a lot of fatty fish, Earth's best store of dietary vitamin D. In 1929 and 1930, Parke-Davis ran ads in *Good Housekeeping* celebrating cod liver oil. "Sunlight that Comes Up from the Sea in Ships," "Golden Gallons of Sunshine from a Wintry Sea," and "Vitamin Farms in the Icy Sea" told of the hardy fishermen who braved the chilly north to secure "nature's greatest substitute for the warm summer sun." Though the tales of tough and heroic outdoorsmen might have been overblown, Parke-Davis's general message was correct: its dietary supplement was nature's best way, short of a sunbath, to secure vitamin D.[26]

By the late 1920s, Squibb's once-small cod liver oil ads had grown large. In national serials and major newspapers, the company told about the health its product gave youngsters. Ads frequently featured a bottle of cod liver oil, now called "bottled sunshine," with light radiating from it. Squibb encouraged mothers to recognize the limits of their own knowledge and trust in medical science: "Your baby may *look* plump and rosy—the picture of health," but an X-ray would reveal otherwise: "Even the healthiest-looking baby is apt to be developing incorrectly in a way his mother never dreams of." Government experts, ads continued, had figured these disturbing facts out and were broadcasting the "grave danger to babies." Generally, the featured risks were skeletal, but mouths full of crooked, decaying, uneven teeth; bent legs; and collapsed chests were shameful deformities in children that would make for a sad adulthood. Only irresponsible parenting would deny children their cod liver oil because "any mother knows how much these things mean to the appearance and the health of her child, not only in babyhood *but* all through life."[27]

Squibb moved beyond motherhood, infants, and rickets, however, when it wrote of the near-miracles its products performed. Light was both healer and powerful metaphor in ads that described the company that carried the "the light of medicine" to overcome "man's ancient enemy, disease." In the late 1920s, Squibb began running a series of promotions hailing the arrival of preventive medicine. Antitoxins had vanquished diphtheria, measles, scarlet fever, and typhoid. Tuberculosis would soon be a memory; rickets endured but not for long. New knowledge about infant feeding would soon end malnutrition. Threatening, old diseases would find no purchase in a world dominated by better medical thinking: "Today, the black death slumbers. Its legions are as terrible as ever, but they march no more through lands where medical science stands on guard." Squibb was vigilant, and because of its work "the white light" of medicine continued to burn: "Brighter and brighter it grows piercing the farther darkness, lighting the way to health and happiness." "The darkness of the middle ages," when pestilence was king, was history; time had brought a new dawn: "The sun of modern science arose. Today the average child born in its light can be expected to live for forty more [years]."[28]

The pharmaceutical company made an even stronger connection between preventive medicine and cod liver oil with its assertion that vitamin D helped protect children from disease. The sunshine vitamin built a "wall of defense" for the sick and well alike. In claims that the AMA would later reject, cod liver

- a well shaped head
- a fine full chest
- sound even teeth
- straight legs
- a strong back

These for your baby with the help of Bottled Sunshine

Figure 5.1. Squibb advertisement, *Ladies' Home Journal*, October 1931, 116.

oil was said to tone the body and hasten recovery from illness; it was a source of "protective, restorative, vitalizing and growth-inducing vitamins!" Some ads mentioned vitamin A alongside D, and few disaggregated the elements: it did not matter what, precisely, in "bottled sunshine" was sunshine. When it was most genuine, Squibb conceded its own ignorance, but the gesture, reminiscent of the half-hearted confessions of many sunlight advocates, had little real consequence. Science was showing the merits of sunshine and of what had become, for many, its friendly substitute.[29]

Squibb found a familiar set of problems in dark America. Babies needed sunshine, and they could get it if not for stifling clothes, bad weather, and ill-designed glass. Fortunately, though, "modern mothers" were aware of the limits of their children's environments. In other ads, the problem was winter. Parke-Davis often privileged the seasonality of sunshine with its ads featuring tough fishermen in the cold, dark months. Another marketing campaign featured a

THE LIGHT THAT SHINES FOR ALL

Figure 5.2. Squibb advertisement, "The Light That Shines for All." Series 11, box 41, N. W. Ayer Advertising Agency Records, Archives Center, National Museum of American History, Smithsonian Institution.

bathing-suit-clad youngster lying in a spoon filled with cod liver oil, the medicine every bit as effective as a summer day: "Sun filled hours of bright, warm beaches . . . eager little bodies tanned a glorious brown. . . . These are precious hours of health which, happily, can be carried over to the cold, dark days of winter." In the 1920s and 1930s, the greatest testament to Squibb's association of its product with winter—and a considerable indicator of its readers' savvy—came from the advertisements it did not produce. The drug company did not even print promotions for its cure-all in *Good Housekeeping* in the summer months, choosing to feature other pharmaceuticals after April and picking up its cod liver oil season each fall.[30]

For children, no doubt little aware of the good cod liver oil was doing in their bodies, it was the product's taste that was most notable. Squibb tried marketing a minty flavor and Coco-Cod, a chocolate alternative. Parke-Davis claimed its product was as tasty as any could be. Apparently, as good as possible was

We sincerely believe that no woman to whose care has been committed the health and well-being of a CHILD .. can pass over this page

SUNLIGHT—*giver of Health . . . In those rays that come hurtling down to us from the sun, millions of miles away, scientists have discovered the source for a vitamin which is essential to the building of sturdy bones, sound teeth, strength and vigor for your children. They have named it Vitamin D—the sunlight vitamin . . .*

Figure 5.3. Cocomalt advertisement, *Ladies' Home Journal*, February 1929, 124.

not good enough for youngsters, and sunshine advocates often paired their celebration of cod liver oil with a realistic critique of its unsavoriness. A wonderful White's Cod Liver Oil Tablets ad, "Cod Liver Oil Nightmare," showed a girl waking up with a start as a giant, terrifying cod attacked her, its teeth bared. White's product—and there were others—was a "pleasant," "candy-coated" alternative to the traditional treatment, a good way to get adults their daily ration, and when mashed up, a palatable dosage for infants.[31]

Certainly, cod in pill form was an improvement, but before long, a new product that was far better would come to dominate the vitamin D market. In 1924, Harry Steenbock at the University of Wisconsin and Alfred Hess at Columbia independently reported that they had discovered that ultraviolet irradiation could enrich foods with vitamin D. The resulting findings, some accurate, others not—cows fed on sunlit grass made vitamin-rich milk, and still more bizarre, air could be D fortified—galvanized the sunlight debate in new ways. Ironically, it was this fortification revolution, which threatened to make sunshine entirely irrelevant, that best distilled the arguments of the time, highlighting innovation that would push forward to a brighter future and putting commerce at the center of the sun discourse.[32]

Though the race to vitamin fortification was roughly a dead heat, it was Steenbock who acted first to secure the rights to his process. Four years after he completed his work making antirachitic food with ultraviolet rays, it received official sanction as patent number 1,680,818. Steenbock's application is informative as much for its bad science as for its good. He pointed out the disturbing commonness of rickets and the unpleasant taste of cod liver oil. He described the process by which an eighth inch of olive oil exposed to a carbon arc or mercury vapor lamp became antirachitic. He saw applications in dog biscuits and chicken feed and thought he could make cereals, grains, seeds, and meat medicinal.[33]

Steenbock's science got more speculative when he tried to figure out how the process worked: "The effect of activating food materials is to cause the activated constituent to emit, in the body, rays, which perhaps are of invisible character, and which, in some manner, cause the calcium depositing cells of the bone to function properly." It appeared, to him, that the activated food actually reemitted rays—the sun was within. Steenbock saw, for his exciting findings, two uses; his process would vastly improve the profitability of existing food products and provide a wonderful service to the health of humanity.[34]

In a noble gesture, Steenbock decided that he should not profit from his findings and tried to donate the patent and any money it would make to the University of Wisconsin. The board of regents did not think administering a patent was the work of a school, so it declined. Eventually, the Wisconsin Alumni Research Foundation (WARF) was established. That body's story has been told elsewhere, and its creation is only of peripheral interest here. It would oversee the Steenbock patent and others, paying commissions to any professor who donated a saleable discovery.[35]

W. A. Evans, the serialized science writer, seized on the story, claiming that Steenbock and Hess's continued "voyages of discovery in the effects of light" had resulted in more breathless anticipation than the adventures of Columbus, Pizarro, and Cortez. According to Evans, by 1929, fortification science had grown more refined. A few years prior, scientists had found that sterols were important to the process and in 1927 had begun work with one in particular, ergosterol. By the end of the decade, 1/1000th of a milligram of irradiated ergosterol was curing rickets in rats, and trials on children were under way. With the new substance, which one report claimed could be 75,000 times more powerful than cod liver oil, the challenge became how to limit, not increase, the potency of vitamin D–fortified products. That led the American Medical Association's Council on Pharmacy and Chemistry to create a standard, viosterol, 100 times as potent as cod liver oil.[36]

The Steenbock patent's effects reached far, with several companies quickly purchasing the rights to fortify. But the Wisconsin researcher did not patent his product for pure profit. He hoped that by limiting those who could use his process, he could ensure that no disreputable companies duped the public. He could also steer industry in minor ways. General Mills got to use the patent, and so did pharmaceuticals like Squibb and Parke-Davis. Milk makers would eventually become the nation's most prominent source of vitamin D (for many reasons, some enumerated later), but margarine producers, competitors to Steenbock's local Wisconsin dairymen did not receive WARF approval.

Predictably Parke-Davis was very interested in this new field of vitamin fortification. The notebook of A. D. Emmett, the company's foremost researcher on D, traces its involvement with the newly discovered process. In 1927, he wrote to Alfred Hess trying to circumvent the impending Steenbock patent approval. Emmett asked the Columbia researcher if he could not make his own claim, having conducted similar research at about the same time. Unfortunately for Parke-Davis, Hess, while disturbed that his finding had been preempted, was unwilling to take up the cause. Still, though Emmett marveled at WARF's ingenious decision to quickly license out Steenbock's process to companies that would then have an incentive to defend their exclusive use of it, he was confident that the patent would collapse under the weight of litigation.[37]

By 1929, Emmett had been proven wrong about the legal onslaught, and Parke-Davis had become one of WARF's guardians. It began producing viosterol 100, then started to contemplate viosterol 250 or 275 or 300. Once the enemy, by 1930, Steenbock was steering internal discussions, and Emmett claimed with confidence that the Wisconsin researcher would oppose anything less than 250. This attempt to standardize dosage and determine the body's needs encountered many challenges, not the least of which were four distinct units of measure. But at least scientists could take pride in their small steps towards agreement: by 1933 there was a conversion system, 2.7 Steenbock units equaled 1 international unit equaled 1 USP unit equaled 3.25 ADMA units. Much as was the case in the rest of the sunlight debate, precise dosage was critical, and that required standardized units of measurement.[38]

Though Steenbock's initial hypothesis, that sunrays were stored in food and released into the body, was inaccurate, he was not alone in his formulation. Both the *Los Angeles Times* and the *Washington Post* reported on discoveries conducted at the Rice Institute that found a new source of genetic mutation: ultraviolet rays trapped inside the body. The sense that plants and animals used energy from the sun as fundamental building blocks for health was far more common. In some ways this was accurate; the sun is indeed a tremendous power source, which gives life to plants, which give life to animals, but not in the literal way early fortification advocates had come to think. In articles, authors told about the "radiant energy" activated foods contained and about the mothers who could wisely bring that force to the table. Predictably, the *Los Angles Times* advocated for California agriculture and played up farm products' ability to trap light. It patronizingly acknowledged that the new fortification experiments and the great successes they promised might seem confusing—so much new information for simple housewives to think about—but in reality, there was little reason for all those intellectual contrivances. Reason dictated that if food was grown in light, food contained light: "At least the desert tomato should contain sunshine if any food does, for it has been coaxing the sun for months." The tomato and the other members of the vegetable world's "welcome triumvirate," cucumbers and onions, joined with desert climate to "do their bit toward keeping up the general good health of our American families." Indeed, another article maintained, "everything that lives is a sun machine," stirred to action by a ball of energy 93 million miles away.[39]

Sunshine Gatherers, a video extolling the wonders of Del Monte fruits, put all this thinking together, telling the complete story of California's "World-Famed orchards" and of sunlit fruit's travel from tree to table. In the film, fruits rest side by side, gathering sunshine in the "veritable Garden of Eden." Women happily can, hard work that ensures the "gathered sunshine is preserved": "Thus, the miracle of canning merges all of this orchard sunshine into one long fruitful season for your table."[40]

Stored sunshine was an attractive notion to people concerned that they too often endured darkness, but fortified-food advocates contended something more complex than that a few tomatoes grown in a field, picked with care, and canned quickly contained the sun's health-giving properties. They claimed, as did so many others, that modern lives carried grave risks and that science could coax a little more goodness out of nature with benevolent "natural" technologies.[41]

Fleischmann's Yeast was great for constipation—or so the company's argument went. Using claims John Harvey Kellogg popularized, Fleischmann's contended that one of the primary problems facing the nation was weakness, and the root of weakness was sluggishness, and the cause of sluggishness was clogged bowels. By 1929, Fleischmann's had identified another enemy of man. In September ads, it claimed that "for lack of sunshine" people were "starved"! They had weak bodies and decaying teeth: "[the] fearful price we pay for the sunless lives we lead." As was the case elsewhere in the sunshine discourse, the roots of darkness were clear. Ads showed a gray and darkened cityscape. Lives were spent in "cave-like offices and factories, in sunless schools, shops, [and] homes" made for a "sunless age." But Fleischmann's offered hope. A banner across the top of a two-page ad in *Good Housekeeping* showed a family basking in sunlight along with the text, "DOWN FROM THE SUN SPEEDS A MYSTERIOUS POWER TO MAKE SICK BODIES VIGOROUS AND WELL." In the greatest discovery in years, a brilliant American scientist had found the answer to many human problems. Irradiated Fleischmann's Yeast brought the "bone-and-tooth hardening health principle of hours in the summer sun." Bodies were being built tougher, with the protection of sunshine vitamin D.[42]

There was a potential tension between the irradiation part of Fleischmann's enthusiasm and the constipation part. Arguments about the latter had couched many of their complaints in antimodern ways. Grown in an unnatural environment, people needed to better manage their bodily processes. Yeast was a way to unclog the body's waste-management system and rid it of toxins. That back-to-nature thinking might have seemed at odds with the widespread irradiation of food—might have, but did not. For Fleischmann's the tension was hardly worth mentioning at all. Confining work indoors had brought "dull, headachy half-health," but this was the "way to quick, *natural* health renewal." It was the "modern way to combat the damaging effects of modern living."[43]

Fleischmann's had competitors from the start. *JAMA* told of a promise for public health in an article announcing the licensing of Quaker to produce vitamin D–rich cereals. As long as commercial enterprises did not take advantage of the public and fortify haphazardly—too much fortification was as problematic as too little—a benefit for health was on the horizon. In its early ads, Quaker latched on to its link with WARF, proudly declaring the partnership and claiming that it was the only cereal that used the "Irradiated Steenbock Process." In one ad, it warned, "Don't be confused. The mere fact that grain is grown in sunlight does not give cereal the 'sun' vitamin." Along with its sister product Muffets, Quaker Farina offered health. It had "a full hour of sunshine in every

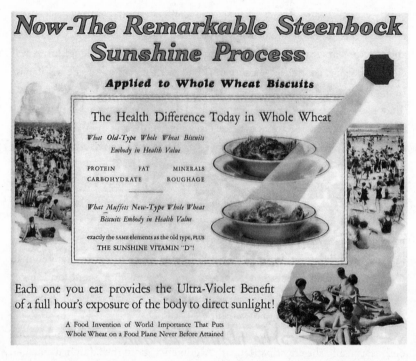

Now-The Remarkable Steenbock Sunshine Process

Applied to Whole Wheat Biscuits

The Health Difference Today in Whole Wheat

What Old-Type Whole Wheat Biscuits Embody in Health Value

PROTEIN   FAT   MINERALS
CARBOHYDRATE   ROUGHAGE

What Muffets New-Type Whole Wheat Biscuits Embody in Health Value

exactly the SAME elements as the old type, PLUS
THE SUNSHINE VITAMIN "D"!

Each one you eat provides the Ultra-Violet Benefit of a full hour's exposure of the body to direct sunlight!

A Food Invention of World Importance That Puts Whole Wheat on a Food Plane Never Before Attained

Figure 5.4. Muffets advertisement, *Ladies' Home Journal*, June 1930, 97.

dish." Ads drove home the point with two pictures, a child eating a bowl of cereal and a child with his bucket at the beach. Using a light thirty times more powerful than sunlight, the company "store[d] up" nature's gift.[44]

It was Bond Bread, however, that most enthusiastically pitched the benefits of the sunshine-grain unity. The baker's ads took several forms, but the healthy, naked child—standing tall with arms raised, pedaling his tricycle, or on roller skates—was common to many of them. Children were naked because that was how they belonged: "If your child went naked in the noon-day sun, any kind of bread might do." Of course, most little ones did not, so for them, Bond Bread was an important alternative. While Fleischmann's often claimed that it offered a product ideal for the young—especially those up to twenty-five years old—Bond would not settle on a target demographic. They told adults of the importance of sunshine acting on skin and encouraged the elderly to rely on a health element that could prevent brittle bones. Bond's story was familiar; it told of a scarce vitamin but an important one and of the growing use of sunlamps and sun-substitutes in medicine. But this new product was, best of all, a healthful, tasty, and economical way to supplement the inadequate provision of good light. The company managed to convey health without souring its baked goods' taste or raising its price: "You can't see sunshine vitamin-D. You can't taste this wonderful health-source any more than you can taste sunshine itself. But it's there—in every loaf of Bond Bread and at no extra price." In one ad, the company recast the merits of its product in a way Saleeby and his fellow eugenists might have appreciated. Bond Bread was a part of the struggle for a well and better nation. In "Mothers, *attention!*" the company called on housewives to safeguard the health, strength, and beauty of children and their "body frame[s]." The request was in a text box in the ad's top right corner entitled "Building an American Citizen."[45]

Excited about its product, the company advertised that it had the approval of WARF, the AMA, the Good Housekeeping Bureau of Foods, the Paediatrics Research Foundation, the Physical Culture Institute, *Child Health Magazine*, *Parents' Magazine*, and the Home-Making Center. For those interested in further information about the company, there was the pamphlet *The Sunshine Trail to Happier Health* and four sponsored radio programs in the New York area, including "Sunshine Health Talks" and "Sunshine Dance Music." Shortly before the first Bond ads came out, a letter from Parke-Davis's Emmett foretold ruin for the baker. It would go the way of the old and unsuccessful vitamin B Ward bread. Emmett thought fortified foods would stumble because products had a peculiar taste and odor. That killed sales. He concluded all such endeavors were doomed; edible vitamin D was destined to be and to remain a pharmaceutical.[46] A. D. Emmett was wrong again. D-fortified bread might never capture a broad market, but there were alternatives.

Dairies had been selling their ability to provide edible sunshine well before they could actually do so in a meaningful way. In fact, there was little vitamin D in milk, but in the late 1920s, companies like Borden's and Blue Valley Butter, perhaps piggybacking on reports that cows raised in light trapped it in their milk, claimed they were bringing the sun to patrons. Ads asking "How Much Sunshine Do You Buy?" and "Are You Darkening Their Lives?" demanded closer attention to the needs of children. Another ad tied a life of competitive disadvantage to a dark childhood. Far from an admonition to relax and slow down, such contentions cut against the grain of so many sunshine assertions. "Life's Race Demands 'Edible Sunshine,'" reminded parents:

> Millions are in this race. No favors asked or given. Success will go to the fittest. A body handicapped by some childhood impairment must fall behind. But

the child whose body has been fortified with the growth and energy elements found in vitamins A and D—all else being equal—will forge to the front.

Arguments presaged the future marriage of milk to vitamin D. At the same time, a few dairies that had already begun to fortify were telling a different—perhaps whiter—lie when they claimed that the little bit of cod liver oil they used did not change the taste of milk.[47]

In the early 1930s, milk makers began fortifying their products in a new way, one that made a healthier beverage without a nasty taste. In time, the D-fortification process became every bit as much a part of milk production as pasteurization and brought the nation into the post-rickets era. There were some good reasons to focus fortification efforts on milk. It made sense to improve a food that was already prominent in children's diets, and since vitamin D required calcium to have its beneficial effects, why not add the substance to a food already rich in the necessary element? Finally, there was a reason that had less to do with children's welfare: WARF and Steenbock's strong ties to statewide dairymen.

In 1932, G. C. Supplee told a group of public health experts that the rise of activated milk was coming. Scientists had perfected a process that permitted the treatment of 5,000 quarts per hour with great uniformity and excellent quality. A year later, *Hygeia* claimed that work had begun, and with Hess's encouragement, Boston, New York, and Philadelphia were fortifying. Papers in Chicago were reporting on the widespread sale of sunshine milk, and before long, both in the Second City and in Gotham, health experts were coming together to outline acceptable standards for vitamin D milk. Consumers could buy it in more than a hundred cities in twenty states.[48]

Many arguments behind the sunlight-milk thinking were not new—pollution, winter, clouds, and cities were all familiar themes. Ads proclaimed that 50 percent of children had rickets; an insidious condition, even "well-fed" and "positively plump" Betty had it. The affliction crippled rich and poor, black and white, and pregnant and nursing mothers too. Its legacy was unenviable: bowed legs, knock knees. Nature did not anticipate or provide for shut-in modern living. Milk was essential for children but also helped adults, bringing them vigor and health. Mothers who gave their girls milk made them beautiful in the present and gave them bright prospects for a comely and happy future: "gleaming, even teeth" in a "beautifully modeled head well poised on shapely shoulders, firm sound flesh nicely distributed over a perfectly proportioned body—plus the bounding energy that comes from perfect health." That youth set the stage for "the glamorous days of courtship, the happily married days, the satisfying days of motherhood." Put simply, sunshine milk would bestow "the glow of health . . . the sunshine glow," the sign of nutrition "vital to your family." "Milk actually helps lengthen life. No stronger evidence can be advanced than the fact that milk-drinking races live longest."[49]

Despite such boosterism, milk ads were at times more cautious in their claims than, for example, Bond Bread. One ad featured a cowboy-hat-wearing, mustachioed patent-medicine salesman with an X through him and an attractive, modern-looking couple pouring a glass of milk. But even with this implied promise of a longer and healthier life, the ad was careful not to overreach: though "the greatest protective food yet discovered," milk was not a cure-all. Perhaps it was because they came later to fortification that dairies emphasized the scientific rigor that brought sunshine into their product. Sealtest provided dairy oversight by certifying that products were bacteria free. By the mid-1930s, the system of laboratories also offered consumers a homogenized vitamin D product, with ads featuring dosage units. According to Chevy Chase, a member of the Sealtest family, each quart of their product contained 400 USP, the equivalent of 150 Steenbock units of a product approved by the AMA. In an ad boasting a similarly precisely fortified product, Sheffield Farms featured a milkman who brought vitamin D when nature could not: he trudged through the rain to a baby who lay atop a calendar for the dark month of October; fortunately, with 400 USP of fortification, "Every Day is *Sun*day for Baby."[50]

But the combination of milk, medicine, and modernity was not always easy. When Sheffield Farms advertised its use of vitamin D extracted from cod liver oil—a process, the company boasted, developed and overseen by Columbia University laboratories—it made a move many other companies were afraid to make. For these other dairy producers, the goal was a natural process. Afraid of precisely the antifortification response Emmett foresaw, they sold their procedures as nature's own—untainted by foul-tasting supplements.

The sun was a common icon in milk advertising, but it was not always easy to argue that fortified milk was no different yet fundamentally different from its light-poor cousin. In the mid-1930s, Pet Milk offered cookbooks that used its product in recipes for tasty treats and healthy meals. It claimed that fortification did not come "by adding some foreign substance to the milk"; therefore, the answer to every good cook's question "Is Pet Milk a substitute for milk?" could be an emphatic, boldfaced no. It was milk, pure and simple. Radiation, just like that from the sun, did nothing more than make a good product a little better.

Not content with this argument, Pet Milk further claimed that it was by drinking fortified milk that people reached a more natural state. Much like a sunlamp brought the sun indoors, this new product was an invaluable substitute for bodies built for sun: "Nature seems to have intended that we should get our supply of vitamin D by having the ultra-violet rays of the sun create this vitamin for us by shining on our bare skin." Unfortunately, the ad explained, "we've changed our way of living since nature made her original plans for us." This had led to vitamin-starved living, but irradiated, Pet Milk was a natural surrogate in a synthetic time.[51]

Borden's produced a promotional pamphlet that aggressively made the same point. In *The Adventures of the Vita-Men: A-B-E and G with Their Leader Vit-Man D*, a crew of helpful superheroes milk cows and deliver a tasty and invigorating product. They bring health to the grandmother, sick person, child, and workingman, keeping them all strong with "Mankind's most nearly perfect food." In recent times, with Steenbock's innovative process, Borden's explains, this great foodstuff has become even better. In one scene, Vita-Man D, strongest of all the Vita-Men, stands before a chalkboard, teaching his pals about the process by which milk becomes sunlit. The diagram shows a lamp shining ultraviolet rays on a thin film of milk. At the top of the chalkboard is written, "In Irradiated Milk Vitamin D Is Created in Nature's Own Way." That natural process was why Borden's tasted so good. Vitamin D milk was simply milk. Carnation boasted similarly, "No foreign substance or taste is added." Sheffield Farms's technique may have been destined for the kind of failure Emmett foresaw, but not Carnation's. His critique had been neutralized.[52]

In *Borden's Presents the Sunshine Makers*, a 1935 animated cartoon, the company promoted its new, sunshine-rich product with a playful critique of modernity. The production starts when a gnome kicks in the doors of four comrades. Off to work, they walk along, stopping to hail "his majesty" the sun. In their tree factory, they turn a crank, which extends a huge mirror. Through a siphon, liquefied sun pours into a Rube Goldberg machine and eventually a bottle, which is loaded onto a cart harnessed to a grasshopper. A gnome delivers milk door-to-door, singing about "the ole golden sunshine."[53]

But all is not well. A dark figure, a black-hatted archer, tries to assassinate the good-hearted little gnome; a firefight ensues, arrows against milk munitions. When a bottle lands on the bad guy—who, the next scene will reveal, is one of a village of smoke lovers—the glass container explodes, dousing him. In the next scene, scores of Smokies are singing in their grim, dark town:[54]

Figure 5.5. "In Irradiated Milk Vitamin D Is Created in Nature's Own Way," 1935. Illustration from Borden's promotional pamphlet *The Adventures of the Vita-Men: A-B-E and G with Their Leader Vit-Man D*. Warshaw Collection of Business Americana–Milk, Archives Center, National Museum of American History, Smithsonian Institution.

We're happy when we're sad,
We're always feeling bad,
How are ya?
Terrible.
That's fine.
We're happy when we're sad.

The struck archer tears into town. Milky light shining from his back chases the other Smokies into their homes, and before long, they are all off to war. Their firepower would probably have scared Saleeby—sprayers filled with pollution from a magical bog. At the border of gnome land, they release their darkening pall. But a bevy of low-tech armaments overmatch them: tree-trunk milk-bottle launchers and dairy-dropping dragonfly bombers. In the end, the gnomes charge the Smokies' village, corral them, and force-feed them at a milk fountain. Suddenly, the bad guys are transformed and sing a new song:

I wanna be happy,
I wanna be glad.
I wanna be happy and glad,
And never again be sad.
I never will cry,
I never will sigh,
'Cause I wanna be glad.[55]

The Sunshine Makers required little analytical work from its audience, wearing its meanings boldly. Borden's had indicted smoke and linked its product to welfare and happiness. It used complex—though low-tech and natural—machines to make sunshine-strong milk. But that product was simply a distilled, bottled natural sunshine, "just the thing to keep you feelin' fine." Borden's also associated its product with the past and with nature by using plants, animals, and mythological figures at peace with their environment: the heroes were gnomes who used trees as weapons and animals as aids, piloting dragonflies not planes, and riding grasshoppers not driving milk trucks.

But all good business ideas are worth fighting for, and the logical marriage of sunlight to milk did not eliminate all fortification competition. In May of 1936, for instance, the *New York Times* "Advertising News and Notes" announced that Schlitz was expanding its campaign—already extensive in the Midwest and New England—for "the beer with sunshine vitamin D." The beverage was better than its competition and "more than refreshing," combining that old Schlitz taste and bouquet with 100 USP of vitamin D and all the summer vitality that it conveyed. The result was "radiant health," straight "from the sun itself." Ads bragged about the product while celebrating the vigor it gave, portraying hustling outfielders, elegant divers, and grinning fishermen.[56]

Though Schlitz relied little on the point, many advertisers were sure to highlight that their products were easy on the wallet. Bond Bread was very clear that its product contained a new health element without sacrificing its usual good taste or low price. Milk makers, too, pointed out that they were offering consumers a healthier version of their old product at no added cost. After stating clearly that the new Pet milk was the same price as the old, a company ad asked, "Can you afford to give your family any other kind of milk?" The answer was, of course, no: "You can't afford to deny your children the protection which this extra vitamin D provides against the faulty development of bones and teeth that may maim them through all their lives." Carnation also made the switch to vitamin D milk an economic no-brainer.[57]

But was that what people really wanted? Were they satisfied with the simple, economical solution dairymen offered? Other milk ads indicated that probably was not the case. The ideal was a vacation for that sun-induced vibrance, but when recreation outside was hard to come by, milk could suffice. In "Keep Your Vacation Vitality!" Deerfoot Farms advertised a nonfortified product, "the aristocrat of milk."[58] "Tanned, healthy and eager, you're back to work . . . the children off to school. For a while, you'll all coast along on your vacation vitality. But winter has a way of wearing down your reserve. Don't let that happen! Keep your vacation vitality—*start drinking milk today*." Clover Dairy and Sheffield Farms sold a similar healthfulness, promising to preserve the fitness captured during a summer outside. Their products were the practical and economical solution to an otherwise untenable indoor life.[59]

Vacation advocates had pointed to a clear problem but would not offer the solution. If life outside was good and leisure the ideal, a nation of white-collar workers and diligent students could not possibly hope to live well. Lamp makers and glaziers claimed that they had democratized sunlight. A $3.95 carbon arc unit could bring the sun cheaply, and cellophane over chicken wire was even less expensive. The matter was not really solved until the arrival of widespread fortification. The ideal might still have been a trip to the beach, a reality even dairy companies recognized, but fortification offered an alternative and far more accessible hope.

This discourse, however, did not break neatly along class lines. Though milk was a good solution for the poor, sun substitutes were not simply for the impoverished. Glassmakers targeted hardworking managers, lamp makers featured glamorous women, and milk companies portrayed the children of successful and modern parents. Rickets afflicted all little ones similarly, and the demands of modern life left both the poor and the well-off susceptible to the diseases of darkness. Financial constraints pushed the ideal of a vacation—or a sunlamp—out of the reach of many families, creating a two-tiered system in which some could afford a couple of weeks of bright vacation, and many could not; but for all Americans, the new realities of work and life meant reliance on sunshine alternatives at least some of the time. Fortunately, innovation had produced a ubiquitous, almost perfect natural healer.

By 1940, the Children's Bureau's *Sunlight for Babies* had become *Substitutes for the Sun*. The concerns were familiar, especially the deficiencies of temperate-zone sunlight and the difficulties of diagnosing minor to moderate rickets. Sunlight was essential in childhood for preventing irritability and malformed bones. But little more than a section in previous pamphlets, dietary substitutes by 1940 merited their own separate publication. This new informational material claimed that, in most of the country, supplements were necessary. Though some babies could grow healthy consuming only 400 USP each day of vitamin D, many needed twice that amount.[60]

According to the Children's Bureau, the solution to America's sunshine woes now came in a bottle. Although in the late 1930s, advertisements for Sheffield Sealect Milk still showed a beach scene or two and explained the importance of milk during wintertime, promotions no longer featured the sun as prominently as they once had. Emphasis had shifted to vitamin D, calcium, and phosphorous. Much as ultraviolet light had become its own element apart from sunshine, vitamin D had become comprehensible to the public as something other than the "sunshine vitamin." Science, with the help of enthusiastic publishers and willing advertisers, had educated the public about its individual merits.[61]

The sunshine story started with the boldest of hopes, a city planned for brightness. It evolved under the watchful eyes of scientists, doctors, educators, and businessmen concerned about the productivity of their workforce. The public caught on. Worried that the modern environment carried grave risks, they sought lifestyles in better sync with what they believed were natural tendencies. Few, however, foresaw a nation clad in loincloths and moccasins. The future would balance old and new, as fortified milk did. People could embrace *simple* machines to add *natural* radiation to regular old milk.

Contrary to the hopes of early tenement reformers who worried about high costs for a vital, natural health element, by 1935, sunlight had emerged as fully saleable, with businessmen advertising their products as sunshine rich. Bond Bread may have claimed it could give health for free and the AMA may have counseled that it could teach the public how to get it cheaply, but both clearly thought healthful sunlight could be pitched to an informed audience. Aware that Americans had become knowledgeable about fortification, companies that made dietary supplements redirected their advertising focus, relying on more scientifically sophisticated assessments of sunlight and vitamin D. Though, over time, USP and Steenbock units began to appear more often in dairy ads than puns about "*Sun*day," the ground had been laid for decades of sunshine enthusiasm and the association of tanning and outdoor recreation with health. Los Angeles developers, the Tucson Sunshine Climate Club, the St. Petersburg newspaper publisher, and the French Lines Sunshine Cruises were counting on just this state of affairs to sell brightness. However desirable, wonderful trips and beautiful vacation homes were little more than dreams for most Americans. They had to rely on an alternative that would ameliorate their pernicious environment. New lifestyles had shaken the country, but the outcome involved a profoundly and predictably American response. Natural good health in this troubling modern world could not be found by looking to an idealized past, for there was no time machine; it came in a bottle, and happiness and welfare were available for a price.

Sunlight into the Twenty-First Century

When John Griscom stepped out of those New York tenements in the middle of the nineteenth century, he spoke of a new American environment. The colonial era had towns, and factory cities like Lowell, Massachusetts, had grown famous in the first half of the nineteenth century, but there had been and was nothing in America like 1840s New York. In the coming years, places like Chicago would mushroom, and buildings would climb skyward. Americans were uncertain about these metropolises, aware of their promise, but afraid that they had begun to sever the tie between people and the natural environment. The new cities were big, they were crowded, and they were polluted. Before long, Americans took action to secure what a generation before they had assumed was a birthright: sunshine.

By the 1920s, the social approaches of city planning and school reform seemed less viable means for meeting the nation's light needs. Recent discoveries, which claimed that sunlight did not penetrate windows and that it struggled to make it through even clear atmosphere—never mind a polluted sky—had created a fresh set of concerns. In this period, sunshine requirements took on a new sense of urgency, and expert scientists and doctors asserted a role in light therapy. Authorities did not limit themselves to sanitarium patients, neither did they presume health came with a haphazard sunbath. They sought better, more scientific ways to bring wellness. God's creation was an inspiration, but scientific thinking prevailed in attempts to secure hypernatural solutions to unnatural ailments.

In the end, the pre-1920 ways of dealing with darkness fell short of their advocates' goals. Housing reformers continued to claim that sunlight mattered, and New York looked to build light into its projects—but big social solutions were out, replaced by individual interventions: personal treatments for sickly patients or sunlamps for sale to modern mothers. In homes, on beaches, and at nudist colonies, Americans were encouraged to proclaim their love for the sun. Many did, and in doing so, often turned their backs on those very medical authorities who spoke openly about the merits of light but hoped to control treatment.

By 1930, the tan was in, and not just as a result of changing tastes. Tall, dark, and handsome had become healthy and vigorous. California, Florida, and the desert Southwest capitalized. So did beaches and lakes closer to northeasterners' homes. But eventually, the health issues related to light deficiencies were solved as much in a glass as at the beach. Vitamin-fortified milk would cure the horrible diseases of darkness. Though milk had largely taken the place of medicinal sunlight, many of the trends in place prior to 1935 continued for some time afterward. Sunlight was worth fifty cents to Jacob Riis's squalid-tenement dwellers, fifty dollars for mothers looking to buy a lamp, hundreds to travelers casting off on Caribbean cruises, and thousands to new California transplants. It remained valuable to vacationers and penthouse purchasers. The South, the West, and the beach, wherever beachgoers found it, continued to capitalize on sunlight's association with vigor, and the bronzed beach bunny and tanned leading man did not fade as ideals for beauty and physical vibrancy. Milk remains vitamin fortified, and the perspective that said science can isolate the best in nature and give that gift to humankind endures. Indeed, instead of resigning themselves to a world devoid of natural light, people found a way to refashion modern environments. This process of making peace with the new by reclaiming the benefits of the old is far from unique to sunlight's story.

But by 1960, another sense of sunlight had begun to emerge. Once celebrated as a source of health and a cure for skin ailments, ultraviolet light began to be seen as more of a hazard than a help. The AMA had long cautioned against searing sunburns, and a handful of doctors hypothesized a role for ultraviolet radiation in the formation of tumors, but few listened. With cancer rates rising—precipitously in the '70s and '80s—those few voices became many, and the public perception of sunlight's role in wellness began to change. In 2002, the National Institute of Environmental Health Sciences added ultraviolet light to its list of known carcinogens; by then, other federal agencies had made skin care and sunlight their business too. The Health and Human Services publication *Healthy People 2010* offered a broad array of prescriptions aimed at preventing a wide variety of diseases. For skin cancer, education was the answer. *Healthy People 2010* sought an increase in the number of people who followed at least one of four prescriptions: stay out of the sun from ten to four, use sunscreen with an SPF of 15 or greater, wear sun-protective clothing, and avoid artificial sources of ultraviolet light.[1]

It appears that public education campaigns have been effective at communicating the danger of sunlight and skin cancer, with Americans now significantly overestimating their chances of contracting the disease.[2] Though it would be easier to comprehend medical conditions that follow predictable paths—from contagion to disease or toxin to tumor—1920s sunlight experts were right: biology is rarely so simple. Two people exposed to the same illness-causing agent will react differently. Perhaps Americans overassess risks because they struggle with this inconsistency; it has certainly made work harder for doctors and epidemiologists. Clear causal mechanisms are elusive, but authorities seem confident that the rising incidence of skin cancer over time is due, in part, to lifestyle changes: more time at the beach has negatively affected health. Rickets rates rose because doctors worried about lack of light and therefore became more likely to diagnose the disease. In the case of skin cancer, doctors may have caused disease rates to rise in a different way: at the behest of experts, celebrants sought the sun's beneficent rays wherever they could, and in their quest for health, found melanoma. Doctors, quite possibly, inadvertently helped cause the skin cancer crisis they now seek to cure.

While ultraviolet radiation became far more of a health menace in the last quarter of the twentieth century than it had been in the first three, skin cancer is not the only story to tell here, and sunlight is far from universally feared. Sunscreen advertisements feature kids sitting under a bright sun and blue skies, not cloistered away from harmful ultraviolet rays. Indeed, in helping beachgoers avoid a tan, these products permit a relaxing, rejuvenating day—with an ultraviolet-less sunbath as part of the experience. Ironically, the opposite is true too: tanning salons endure because they offer prized golden skin to patrons who cannot make it to the beach at all. Cosmetic companies claim to have found a happy medium with creams that sell the look of a day in the sun but without the sunlight. A 2004 *Consumer Reports* article detailed tests on seven spray tanners: "They can make you look as if you've been basking at the beach without the worry about skin damage, or the expense of a trip to the tropics."[3]

While consensus holds that sunlight is bad for health, medical experts have not reached unanimity. In fact, if anything, evidence has started to swing in favor of more vitamin D. Harvard epidemiologist Ed Giovannucci caused quite a stir at a meeting of the American Association for Cancer Research when he suggested that some sunlight might do more good than harm. According to Giovannucci, fewer Americans would suffer from osteoporosis and fractures if they got more of the sunshine vitamin, and a myriad cancers that vitamin D might help prevent—of the colon, pancreas, prostate, esophagus, stomach, and breast—together are far more common and deadly than most forms of skin cancer. Giovannucci was not looking to create a nation of sunbathers; he simply suggested that it was time to consider a few minutes of sunscreen-less sunshine because even fortified foods give far less D than ten minutes outside.[4]

The truth, however, is that although Giovannucci may have been bold to ring the changes on sunlight in a room full of cancer experts, he is not alone. In the late 1990s, doctors began to report a rise in rickets rates, especially in the African American community. Since then, doctors have added other concerns about sunlight deficiencies, postulating a role for vitamin D in stimulating the immune system and in heart health and diabetes prevention. Doctors admit that they have lots of studying to do and that they remain uncertain about how much vitamin D people need, but it seems likely that the standards long in place are simply too low and that Americans' health has suffered as a result. Generally, scientists talk about revising recommended doses of dietary supplements, but some also confess that a touch of sunlight might be a way to improve health.[5]

To many Americans, the greatest threat that darkness poses is not physical at all. Seasonal-Affective Disorder (SAD), a form of depression brought on by weak sunlight and short days, attained prominence in the 1980s after a few medical studies and a surge in media attention. Since then, the American Psychiatric Association has come to recognize the role of dark winters in facilitating seasonal, often crippling, depressive episodes. Less severe cases of seasonal mood swings are, no doubt, even more common. The treatment for this troubling condition is reminiscent of 1930s products: light boxes with very bright, broad-spectrum emissions that simulate nature in an attempt to return to Americans the visible light they lose in wintertime.[6]

Parans, a Swedish company, promises an even truer faux-nature. For a substantial fee, the innovative illuminator will install solar collecting panels on buildings, transmitting natural light by fiber optic cable to ceiling-mounted lamps. Parans contends that daylight is good for health and far better for happiness than conventional bulbs, setting essential biorhythms and regulating hormonal secretions. According to the company, and many authorities on SAD agree, the earlier attention to wavelengths was not entirely misguided. The pineal gland keeps people feeling right, and waves at 460 mμ, relatively scarce in electric light, are responsible for helping to regulate pineal secretions.[7]

The Parans installation cannot be had cheaply, but the premium consumers are willing to pay for sunlight is nowhere greater than in real estate. There are few selling points, aside from basics like space and location, more valuable than good, natural light. While it is often hard to quantify the full extent of that premium, a handful of installations make clear the price urban Americans will pay for sunshine. In 2005, the *New York Times* reported on a new setup intended to bring light to a small parcel of Lower Manhattan. The Battery Park City Authority had earmarked $335,000 for three brand-new heliostats—mirrors that follow the sun through the sky and redirect its light—to be installed on the twenty-third floor of an adjacent building. Carpenter Norris Consulting, one of the project's planners, conceded that the installation would not make everything in Battery Park bloom, but that did not discourage James Gill, chairman of the Authority, who, confident that the nearby apartments would increase in value, said, "There's a lot to be said for light."[8]

Carpenter Norris, which lauds the "valuable role daylight plays in a productive workplace," has made a business out of innovative daylighting solutions. Its installation in a law firm office in Washington, DC, is right out of the 1930s. A rooftop heliostat reflects light into a 36-meter-long column, or "solar light pipe," which transmits the sun's rays into the bowels of the building. In 2003, the *Wall Street Journal* gave an account of the installation in the context of many other clever projects that have employed innovative design solutions to bring natural light to buildings. The article generally attributed the popularity of this work to increased energy costs and legislative incentives for conservation, but it prefaced the argument with more far-reaching observations about the irritating buzz of overhead fluorescent bulbs and the long search for better indoor lighting: "Scientists and building designers have talked for decades about the virtues of light from our nearest star. Countless studies show that access to the sun's rays bump up sales in stores, lubricate learning in classrooms and increase productivity and improve health at work."[9]

The town of Rattenberg, Austria, flirted with an installation that would have made Carpenter Norris's work seem insubstantial. The tiny town sits at the foot of two mountains and is therefore shaded from late fall to midwinter every year. For a time, the mayor hoped that he could reverse the dark history of the town's location, chosen six hundred years ago because it would be easy to defend. With help from the EU, Rattenberg planned to install heliostats a quarter of a mile away that would banish dark days from the town's streets. According to an engineer at Bartenbach Light Laboratory, the firm responsible for the plan, the idea went beyond simply lighting the village: "The idea is to give them the impression they have sun." Unfortunately for the mayor and his supporters, high costs and opposition undid his bold plan.[10]

With the Rattenberg solution and the proposed heliostats for New York's buildings and downtown parks, this book comes full circle. If skin cancer indicates a changing sense of sunlight, these solutions to urban problems demonstrate continuity. They show that people know quite well the difference between real sunlight and the alternative a humming fluorescent unit provides. They also show that innovation continues to enable radical solutions to environmental problems—in the case of Rattenberg, a condition brought about by the very location of the town. Finally, the difference between New York and the Austrian valley town dramatizes the Progressive failure. New York's recent architectural innovations brought sunlight to a handful of expensive buildings and promised it for a small park in a high-priced part of town. Certainly, heliostats like Rattenberg's could not bring light to a world metropolis, but the goal of a little sunlight for everyone harkens back to the Progressive optimism of early twentieth-century New York.

That reform movement came at a time when darkness had emerged as a major concern. Though little of those early fears remain, humans have an enduring sense that sunlight is necessary for wellbeing. Moreover, the attempt to combat darkness and the success of innovation in chasing away a dreaded illness of sunlight deficiency offers instructive models when considering other efforts to secure natural advantages and transform the environment. But this is also, to some extent, a cautionary tale. Perhaps doctors and public health officials erred when they counseled the goal of a deep tan and decided that they could replicate sunlight's value with minutes-long treatments from high power, shortwave ultraviolet light; or children and parents missed out when experts concluded that milk was an adequate substitute for a day outside; or city dwellers suffered when architects decided a high-rise city was profitable and planners determined that a lower-rise one was not worth preserving. The story of American sunlight is one of powerful new fears, of remarkable technological mastery, of created commodities, and of hypernatural but limited solutions to environmental hazards. In short, it is the story of modernity's promise and its danger.

NOTES

Introduction

1. Vitruvius, *Ten Books on Architecture*, trans. Ingrid D. Rowland (Cambridge: Cambridge University Press, 1999).

2. T. Bendyshe, "On the Anthropology of Linnaeus, 1735–1776," *Memoirs Read before the Anthropological Society of London* 1 (1863–64): 421–58; Armand Marie Leroi, *Mutants: On Genetic Variety and the Human Body* (New York: Penguin Books, 2005), 247–50.

3. In his book *Rise and Shine: Sunlight, Technology and Health* (Oxford: Berg, 2007), Simon Carter writes of British enthusiasm for sunlight therapy; indeed, this enthusiasm ran deep and accorded with many characteristics in American thinking. Carter's book shares many subjects with this one, but his argument and presentation of the material are quite different. It is, nevertheless, an important contribution to the field.

4. Historians have written a lot on the commodification of resources, including land, water, and air. See William Cronon, *Nature's Metropolis: Chicago and the Great West* (New York: W. W. Norton and Co., 1991); Donald Worster, *Dust Bowl: The Southern Plains in the 1930s* (Oxford: Oxford University Press, 1982); Theodore Steinberg, *Slide Mountain, or The Folly of Owning Nature* (Berkeley: University of California Press, 1996); Donald Worster, *Rivers of Empire: Water, Aridity, and the Growth of the American West* (New York City: Oxford University Press, 1992). Thanks to Gabi Azevedo, who

took the time and expended the critical energy necessary to help me think these arguments through.

5. Gail Cooper's *Air Conditioning America: Engineers and the Controlled Environment, 1900–1960* (Baltimore: Johns Hopkins University Press, 1998) makes a similar argument about the rise of climate control in America, though engineers play a more central role in her work than they do in mine.

6. Cronon's discussion of second nature (see *Nature's Metropolis*) pushes in this direction, but other scholarship is more clearly about similar themes. See Thomas Bender, *Towards an Urban Vision: Ideas and Institutions in Nineteenth Century America* (Baltimore: Johns Hopkins University Press, 1987), and Leo Marx, *The Machine in the Garden: Technology and the Pastoral Ideal in America* (New York: Oxford University Press, 1964). David Nye's *America as Second Creation: Technology and Narratives of New Beginnings* (Cambridge: MIT Press, 2004) tells about the ways that people tried to make their transformations of the environment appear natural. In some ways, my reading of the historical record has even more in common with Michael Pollan's "nutritionism," best outlined in *In Defense of Food: An Eater's Manifesto* (New York: Penguin Press, 2008). Pollan writes of the attempts of food scientists to break down a healthy diet into its constituent nutritional elements and synthesize them.

When it comes to modern/antimodern conflicts, many scholars have pitted modernists against antimodernists or talked about the development of a new twentieth-century sensibility. Sometimes revisionists have discussed the ways that critics of modernity unintentionally reinforced new attitudes. For examples of previous scholarship, see T. J. Jackson Lears, *No Place of Grace: Antimodernism and the Transformation of American Culture, 1880–1920* (New York: Pantheon Books, 1981); Lynn Dumenil, *The Modern Temper: American Culture and Society in the 1920s* (New York: Hill and Wang, 1995); William Leach, *Land of Desire: Merchants, Power and the Rise of a New American* (New York City: Pantheon Books, 1993); Richard Fox Wightman and T. J. Jackson Lears, introduction to *The Culture of Consumption: Critical Essays in American History, 1880–1980*, ed. Richard Fox Wightman and T. J. Jackson Lears (New York City: Pantheon Books, 1983), and in the same volume, Lears's "From Salvation to Self Realization: Advertising and the Therapeutic Roots of the Consumer Culture, 1880–1930"; Harvey Green, *Fit for America: Health, Fitness, and American Society* (New York City: Pantheon Books, 1986); T. J. Jackson Lears, "American Advertising and the Reconstruction of the Body, 1880–1930," in *Fitness in American Culture: Images of Health, Sport, and the Body, 1830–1940*, ed. Kathryn Grover (Amherst: University of Massachusetts Press, 1989).

Chapter One

1. John H. Griscom, *The Sanitary Condition of the Laboring Class of New York with Suggestions for Its Improvement* (New York: Arno Press, 1845), 8.

2. Historians have written extensively on this law and the maneuvering that enabled its passage. See especially Roy Lubove, *Progressives and the Slums: Tenement House Reform in New York City, 1890–1917* (Pittsburgh: University of Pittsburgh Press, 1962), and Anthony Jackson, *A Place Called Home: A History of Low-Cost Housing in Manhattan* (Cambridge: MIT Press, 1976).

3. The literature on Progressive reform is considerable. Robert Wiebe's *Search for Order, 1877–1920* (New York: Hill and Wang, 1967) and Richard Hofstadter's *Age of Reform: From Bryant to FDR* (New York City: Knopf, 1955) hold up well over time. I have also tried to keep Daniel Rodgers's international corrective in mind; see *Atlantic Crossings: Social Politics in a Pro-*

gressive Age (Cambridge: Belknap Press of Harvard University Press, 1998). Indeed, ideas and influences moved across the Atlantic, with American housing, planning, and sanitary reformers learning from Europe. The international exchange of ideas plays an even more prominent role in the subsequent chapters of this book.

4. *Report of the Council of Hygiene and Public Health of the Citizen's Association of New York upon the Sanitary Condition of the City* (New York: D. Appleton and Co., 1865), 103–4.

5. Lawrence Veiller, "Habitable Tenement Houses," *New York Times*, 29 June 1899, 6; Tenement House Committee, "Tenement House Ordinances," *Charities* 3 (1899): 2–6; Ernest Flagg, "New York Tenement House Evil and Its Cure," *Scribner's Magazine* 16 (1894): 370–92; Robert W. DeForest and Lawrence Veiller, "The Tenement House Problem," in *The Tenement House Problem*, ed. Robert W. DeForest and Lawrence Veiller (New York: Macmillan Co., 1970), 3–68.

6. I refer to the following essays from DeForest and Veiller's *Tenement House Problem*: Lawrence Veiller's "Tenement House Reform in New York City, 1834–1900," 113, and his "Back to Back Tenements," 295, 297; Robert W. DeForest's introduction, "Tenement Reform in New York since 1901," xiii–xxxi.

7. *New York Times*, 19 December 1896, 4; "High Building Problem," *New York Times*, 16 August 1895, 9; "Limit for High Buildings," *New York Times*, 5 April 1894, 1; Herbert S. Swan and George W. Tuttle, *Planning Sunlight Cities* (New York: American City Pamphlets, 1917).

8. *Report of the Heights of Buildings Commission to the Committee on the Height, Size, and Arrangement of Buildings of the Board of Estimate and Apportionment of the City of New York* (New York: M. B. Brown Printing and Binding Co., 1913), 1; Commission on Building Districts and Restrictions, *Final Report* (New York: Committee on the City Plan, 1916).

9. Henry B. Fuller, *The Cliff-Dwellers* (Ridgewood: Gregg Press, 1968), introduction.

10. David Stradling, *Smokestacks and Progressives: Environmentalists, Engineers, and Air Quality in America, 1881–1951* (Baltimore: Johns Hopkins University Press, 1999), especially chapter 1; Joel Tarr, *The Search for the Perfect Sink: Urban Pollution in Historical Perspective* (Akron: University of Akron Press, 1996); Martin V. Melosi, "Environmental Crisis in the City: The Relationship between Industrialization and Urban Pollution," in *Pollution and Reform in American Cities, 1870–1930*, ed. Martin V. Melosi (Austin: University of Texas Press, 1980); Dale Grinder, "The Battle for Clean Air: The Smoke Problem in Post-Civil War America," in Melosi, *Pollution and Reform*.

11. Matthew Hale Smith, *Sunshine and Shadow in New York* (Hartford: J. B. Burr and Co., 1868); Mary A. Livermore, *The Story of My Life, or The Sunshine and Shadow of Seventy Years* (Hartford: A. D. Worthington and Co., 1898); Paul Laurence Dunbar, *Lyrics of Sunshine and Shadow* (New York: Dodd, Mead and Co., 1905); Isaac D. Williams. "Sunshine and Shadow of Slave Life: Reminiscences as Told by Isaac D. Williams to 'Tege,'" in the digital collection Documenting the American South, University Library, University of North Carolina, Chapel Hill, http://docsouth.unc.edu/neh/iwilliams/iwilliams.html (accessed 8 August 2003). See also Flora Beal Shelton, *Sunshine and Shadow on the Tibetan Border* (Cincinnati: Foreign Christian Missionary Society, 1912); "Sunshine and Shadow of Stage," *Chicago Daily Tribune*, 15 November 1896, 35; "Before the Emerald Club," *Washington Post*, 5 October 1891, 1. For secondary sources on biases, see Christine Stansell, *City of Women: Sex and Class in New York, 1789–1860* (Urbana: University of Illinois Press, 1987); Peter George Buckley, "To the Opera House: Culture and Society in New York City, 1820–1860" (PhD diss., State University of New York, Stony Brook, 1984).

12. Robert W. DeForest, "Tenement House Regulation—the Reasons for It—Its Proper

Limitations," *Annals of the American Academy of Political and Social Science* 20 (1902): 86; Ernest Kent Coulter, *The Children in the Shadow* (College Park: McGrath Publishing Co., 1969), fore-word, chapter 3.

13. *Report of the Tenement House Committee of 1894 as Authorized by Chapter 479 of the Laws of 1894,* (Albany: J. B. Lyon, State Printer, 1895); Tenement House Committee, "Tenement House Ordinances," 3; DeForest and Veiller, "The Tenement House Problem," 10.

14. Jacob Riis, photo 90.13.1.122, Jacob Riis Collection, Museum of the City of New York. For an example of photographs in alleys, see photo 90.13.1.116.

15. Photographs are, of course, far from new to historical inquiry. In their studies, Peter Con-rad and Alan Trachtenburg have looked at subjects and employed a methodology similar to mine. For both, the picture is not just an indication of the artist's intent; it is also a document of a time, readable as an opaque "mass of facts." See Alan Trachtenberg, *Reading American Photo-graphs: Images as History, Mathew Brady to Walker Evans* (Toronto: Hill and Wang, 1989), espe-cially xv; Peter Conrad, *The Art of the City: Views and Versions of New York* (New York: Oxford University Press, 1984). Peter Hales's *Silver Cities: The Photography of Urbanization, 1839–1915* (Philadelphia: Temple University Press, 1984) treats some content similar to mine but with a far different focus. He looks at reform pictures and traces the technological and aesthetic shifts that converted urban photography into a mass producible form that could reveal a dynamic city in its full complexity.

16. Robley Dunglison, *Human Health, or The Influence and Locality; Change of Air and Cli-mate; Seasons; Food Clothing; Bathing and Mineral Springs; Exercise; Sleep; Corporeal and Intellec-tual Pursuits, &c. &c. on Healthy Man; Constitution Elements of Hygiene* (Philadelphia: Lean and Blanchard, 1844), 53; Edward B. Foote Jr., *The Blue Glass Cure: How and When It Originated; Why It Has Been Ridiculed; Gen. Pleasonton Not a Success as an Experimental Philosopher; His Facts and Theories; What They Are Worth; the Light of Science Brought to Shine on Blue Glass; Panes Cur-ing Pains; Real Merits Made Plain, and the Cob-Webs Brushed Away; the Sun as the Source of All Power; Its Beneficial and Its Baneful Influences; Absence of Sun-Light Injurious; Why It Is Avoided; the Properties of Sun-Light; How They Are modified by Blue Glass; a Few Practical Hints; Etc.* (New York: Murray Hill Publishing Co., 1880), 37; Forbes Winslow, *Light: Its Influence on Life and Health* (New York: Moorhead, Simpson and Bond, 1868).

17. For a solid account of the use of light in medicine in the nineteenth century, see Tanya Sheehan, "Doctor Photo: The Cultural Authority of Portrait Photography as Medicine in Nineteenth-Century America" (PhD diss., Brown University, 2005). General A. J. Pleason-ton, *The Influence of the Blue Ray of the Sunlight and of the Colour of the Sky, in Developing Animal and Vegetable Life; in Arresting Disease, and in Restoring Health in Acute and Chronic Disorders to Human and Domestic Animals* (Philadelphia: Claxton, Remsen and Haffelfinger, 1877); Edwin D. Babbitt, *The Principles of Light and Color: Including among Other Things the Harmonic Laws of the Universe, the Etherio-Atomic Philosophy of Force, Chromo Chemistry, Chromo Therapeutics, and the General Philosophy of the Fine Forces, Together with Numerous Discoveries and Practical Applications* (New York: Babbitt and Co., 1878).

18. Nancy Tomes, *The Gospel of Germs: Men, Women, and the Microbe in American Life* (Cam-bridge: Harvard University Press, 1998), especially part 1.

19. Ironically, blue light, long ago rejected by the medical establishment, has recently emerged as a potential weapon in the fight against MRSA. See Chukuka S. Enwemeka et al., "Blue 470-nm Light Kills Methicillin-Resistant *Staphylococcus aureus* (MRSA) *in Vitro,*" *Photo-medicine and Laser Surgery* 27 (2009): 221–26.

20. Herbert Upham Williams, *A Manual of Bacteriology*, 2nd ed. (Philadelphia: P. Blakiston's Son and Co., 1901); Philip Hanson Hiss and Hans Zinsser, *A Text-Book of Bacteriology*, 2nd ed. (New York: D. Appleton and Co., 1914); Percy Frankland, *Bacteria in Daily Life* (London: Longmans Green and Co., 1903); Arthur Downes and Thomas P. Blunt, "Researches on the Effect of Light upon Bacteria and Other Organisms," *Proceedings of the Royal Society of London* 26 (1877): 488–500; Arthur Downes and Thomas P. Blunt, "On the Influence of Light upon Protoplasm," *Proceedings of the Royal Society of London* 28 (1878): 199–212; Edgar M. Crookshank, *A Textbook of Bacteriology Including the Etiology and Prevention of Infective Diseases and a Short Account of Yeasts and Molds, Hematozoa and Psorosperms*, 4th ed. (Philadelphia: W. B. Saunders, 1897); "Koch on Bacteriology," *Science* 16 (1890): 169–70.

21. John F. J. Sykes, *Public Health and Housing: The Influence of the Dwelling upon the Health in Relation to the Changing Style of Habitation* (London: P. S. King and Son, 1901), DeForest, "Tenement Reform in New York since 1901," xvii; Hermann M. Biggs, "Tuberculosis and the Tenement House Problem," in DeForest and Veiller, *Tenement House Problem*, 447–58; "Light in Tenements," *New York Times*, 25 June 1895, 13.

22. Arthur R. Guerard, "The Relation of Tuberculosis to the Tenement House Problem," in DeForest and Veiller, *Tenement House Problem*, 461–70.

23. William Atkinson, *The Orientation of Buildings, or Planning for Sunlight* (New York: John Wiley and Sons, 1912); Commission on Building Districts and Restrictions, *Final Report*, 9.

24. "Tenement Evils as Seen by the Tenants," in DeForest and Veiller, *Tenement House Problem*, 385–417; Jacob A. Riis, *The Battle with the Slum* (New York: Macmillan Co., 1902).

25. Commission on Building Districts and Restrictions, *Final Report*, 25–27, 95, 105, 109, 151, 166–70, 195–97; *Report of the Heights of Buildings Commission*, 19, 57–58.

26. Lawrence Veiller, "Housing Conditions and Tenement Laws in Leading American Cities," in DeForest and Veiller, *Tenement House Problem*, 131–56; E. R. L. Gould, *The Housing of Working People: Special Report of the Commissioner of Labor* (Washington, DC: Government Printing Office, 1895); Winthrop E. Dwight, "Housing Conditions and Tenement Laws in Leading European Cities," in DeForest and Veiller, *Tenement House Problem*, 173–90.

27. Alfred Tredway White, "Better Homes for Workingmen" (paper presented at the Twelfth National Conference of Charities, Washington, DC, June 1885); E. R. L. Gould, "The Housing Problem," *Municipal Affairs* 3 (1899): 108–31; Gould, *Housing of Working People*; Francis R. Cope Jr., "Tenement House Reform: Its Practical Results in the 'Battle Row' District, New York," *American Journal of Sociology* 7 (1901): 331–58; *Improved Dwellings for the Working Classes* (New York: G. P. Putnam's Sons, 1879).

28. "The Proposed Code of Tenement House Laws," appendix 2 in DeForest and Veiller, *Tenement House Problem*; see also appendix 5, "The Tenement House Act as Amended in 1901, 1902, and 1903."

29. Jacob A. Riis, "The Story of the Slum," *Chicago Daily Tribune*, 11 March 1900, 37 (part 1); 19 March 1900, 37 (part 2); 25 March 1900, 45 (part 3).

30. Robert Hunter, *Tenement Conditions in Chicago: Report by the Investigating Committee of the City Homes Association* (Chicago: City Homes Association, 1901), 14; Jane Addams, "The Housing Problem in Chicago," *Annals of the American Academy of Political and Social Science*, 20 (1902): 99–103.

31. Bonnett v. Vallier, 136 Wis. 193 (1908).

32. Ford H. MacGregor, "Tenement House Legislation State and Local," in *Comparative Legislation Bulletin* (Madison: Wisconsin Library Commission, 1909). See also W. Locke Rockwell,

The Tenement-House Act of New Jersey (Newark, NJ: Soney and Sage, 1904); Pennsylvania, *Tenement House Act for Cities of the First Class* (1895).

33. Emily Wayland Dinwiddie, *Housing Conditions in Philadelphia* (Philadelphia: Octavia Hill Association, 1904), available online at Google Books (accessed 8 August 2010); Charles Frederick Weller and Eugenia Winston Weller, *Neglected Neighbors: Stories of Life in the Alleys, Tenements and Shanties of the Nation's Capital* (Philadelphia: John C. Winston Co., 1908); Mary Buell Sayles, "Housing Conditions in Jersey City," *Annals of the American Academy of Political and Social Science*, supplement 16 (1903); Lawrence Veiller, *The National Housing Association* (New York, 1910); Charlotte Rumbold, *Housing Conditions in St. Louis* (St. Louis: Civic League of St. Louis, 1908); Janet E. Kemp, *Report of the Tenement House Commission of Louisville* (1909).

34. The literature on city planning is considerable. Among this work, Jon Peterson's book *The Birth of City Planning in the United States, 1840–1917* (Baltimore: Johns Hopkins University Press, 2003) about the emergence of a formed field with an expansive vision is an excellent resource and overview. In *Building Gotham: Civic Culture and Public Policy in New York City, 1898–1931* (Baltimore: Johns Hopkins University Press, 2002), Keith Revell offers a strong account of the many ways that New York's planners came together to make a livable city. See also Robert M. Fogelson, *Downtown: Its Rise and Fall, 1880–1950* (New Haven: Yale University Press, 2001), and Carol Willis, *Form Follows Finance: Skyscrapers and Skylines in New York and Chicago* (New York City: Princeton Architectural Press, 1995).

35. "In Favor of Limitation," *Chicago Daily Tribune*, 15 October 1891, 9; "There Should Be Some Limit," *Chicago Daily Tribune*, 13 October 1891, 1; "Problem of the Sky-Scrapers," *Chicago Daily Tribune*, 15 October 1891, 9; "Height of Buildings Limited," *Chicago Daily Tribune*, 13 October 1891, 1.

36. "High Building Problem," 9; Ernest Flagg, "The Dangers of High Buildings," *Cosmopolitan*, May 1896, 70, 79; "Skyscraper Cut Short," *New York Times*, 27 January 1904, 16; "Topics of the Times," *New York Times*, 14 May 1903, 8. See also "Chief of Big Skyscraper Would Curb Heights," *New York Times*, 13 February 1916, SM18. The end to voluntarism is one of Revell's main points in *Building Gotham*. Not everyone was so dismayed by the aesthetics of a skyscraper city. Historians David E. Nye and William R. Taylor have written about the ways New Yorkers made peace with or celebrated their new buildings. See William Taylor, "Psyching Out the City," in *Uprooted Americans: Essays to Honor Oscar Handlin*, ed. Richard L. Bushman (Boston: Little, Brown and Co., 1979), 245–87, and David E. Nye, "The Sublime and the Skyline: The New York Skyscraper," in *The American Skyscraper: Cultural Histories*, ed. Roberta Moudry (Cambridge: Cambridge University Press, 2005), 255–69. In an article, Revell takes the argument one step further, suggesting that the zoning of New York was largely an attempt to direct Gotham's skyline to a more ideal aesthetic. Concerns about public health were secondary, largely grafted onto the law as a justification in order to gain legal sanction. According to Revell, George B. Ford, a leading planner, reframed a concern for "visual unity and harmony" as a concern for "the common stock of light and air." While Revell adeptly shows that planners were deeply concerned with the look of their cities, my evidence calls his analysis into question. Public health concerns were more influential than he suggests, and planners were undoubtedly more worried about the city's environment. See Keith D. Revell, "Law Makes Order: The Search for Ensemble in the Skyscraper City, 1890–1930," in Moudry, *The American Skyscraper*, 38–62.

37. *Report of the Heights of Buildings Commission*, appendices 3 and 4.

38. Rodgers, *Atlantic Crossings*; Commission on Building Districts and Restrictions, *Final Report*.

39. Edward M. Bassett, "Constitutional Limitations on City Planning Powers" (paper presented at the Ninth National Conference on City Planning, Kansas City, MO, 7–9 May 1917), 199–227.

40. *A Standard State Zoning Enabling Act* (Washington, DC: Government Printing Office, 1926); James M. Clark, "The Pittsburgh Zoning Ordinance" (paper presented at the Thirteenth National Conference on City Planning, Pittsburgh, PA, 9–11 May 1921), 155.

41. *A Comprehensive City Plan for Memphis Tennessee* (City Plan Commission, 1924); *Ten Years' Progress on the City Plan of St. Louis, 1916–1926* (City Plan Commission, 1927); *Report of the City Planning Commission with Proposed Zone Plan for San Francisco* (1920); *Report of the City Planning Commission, Houston, Texas* (Forum for Civics, 1929); *A City Plan for Springfield, Mass* (Progress Report by the Planning Board, 1923); *City Plan for El Paso, Texas* (Published by authority of the Mayor and City Council, 1925); Frank Backus Williams, *Akron and Its Planning Law* (Akron Chamber of Commerce, 1919).

42. *Zoning for Milwaukee* (Tentative Report of the Board of Public Land Commissioners, 1920).

43. *Proposed Building Zones for Newark* (Tentative Report of the Commission on Building Districts and Restrictions, 1919).

44. Building Inspector et al. v. McInerney, 47 Wyo. 258 (1934); Emma Sundeen v. James A. Rogers, 83 NH 253 (1928); R. B. Construction Company v. Howard W. Jackson, 152 Md. 671 (1927); Charles H. Bebb et al. v. F. M. Jordan, 111 Wash. 73 (1920).

45. John A. Kingsbury, "A New Experiment in the Treatment of Combined Poverty and Tuberculosis," and Kingsbury, "A Home Hospital: An Experiment in Home Treatment for Tuberculosis," box 45, folder 301, Community Service Society, Columbia University Rare Books and Manuscripts Library (CU RBML). Donald B. Armstrong, "The Home Hospital Experiment," box 45, folder 301-8, Community Service Society. Publication 116 does not provide a year, but it was almost certainly from the mid–nineteen teens.

46. "Welcome Oases in New York's Desert of Rooftops," *New York Times*, 14 August 1910, X5; "Hospital's Strong Ally an Open Air Roof Ward: Experiment at the Presbyterian Institution a Success," *New York Times*, 9 December 1906, 10; "New Mt. Sinai Hospital," *New York Times*, 28 February 1904, 24; "Cost of New Bellevue Will Be $11,000,000," *New York Times*, 23 April 1904, 9; "Municipal Hospital Planned for Care of Consumptives," *New York Times*, 15 March 1903, 28. Sheila Rothman, *Living in the Shadow of Death: Tuberculosis and the Social Experience of Illness in American History* (New York City: Basic Books, 1994); Emily Abel, *Tuberculosis and the Politics of Exclusion: A History of Public Health and Migration in Los Angeles* (New Brunswick: Rutgers University Press, 2007); Emily Abel, *Suffering in the Land of Sunshine: A Los Angeles Illness Narrative* (New Brunswick: Rutgers University Press, 2006); Gregg Mitman, "Geographies of Hope: Mining the Frontiers of Health in Denver and Beyond, 1870–1965," *Osiris* 19 (2004): 93–111.

47. James C. Whorton, *Crusaders for Fitness: The History of American Public Health Reformers* (Princeton: Princeton University Press, 1982); J. H. Kellogg, *Light Therapeutics: A Practical Manual of Phototherapy for the Student and the Practitioner* (Battle Creek: Good Health Publishing Co., 1910). See also Wilhelm Winternitz and B. Buxbaum, *Hydrotherapy, Thermotherapy, Heliotherapy, and Phototherapy*, vol. 9 of *A System of Physiologic Therapeutics: A Practical Exposition of the Methods, Other than Druggiving, Useful in the Prevention of Disease and in the Treatment of the Sick*, ed. Solomon Solis Cohen (Philadelphia: Blakiston's Son and Co., 1902), supplemental chapter 1. For more on sanitaria and their history, see chapter 3.

48. Constance D'Arcy McKay, "Pageant of Sunshine and Shadow," box 2, folder 5, Constance D'Arcy McKay Papers, New York Public Library Performing Arts Library. Lewis Hine, "The Dance of the Children of the Sunshine," Prints and Photographs Online Catalog, National Child Labor Committee Collection, Library of Congress, http://www.loc.gov/pictures/item/ncl2004004469/PP/ (accessed 12 August 2010).

49. Compiled from the New York City Department of Education Annual Reports, Records of the Department of Education, New York City Municipal Archives.

50. It is far harder to get a precise number of H-shaped schools than of rooftop playgrounds because a few of the buildings that appeared H-shaped in this period were probably U-shaped (desirable to Snyder for similar reasons). This estimate, compiled from architectural drawings and photographs housed in the New York Municipal Archives seems on the conservative side. Jean Arrington has conducted a survey of all of Snyder's buildings and turned up a total of forty-six H-shaped buildings, with thirty-five still extant (e-mail correspondence from Jean Arrington, 6 October 2009). Riis, *Battle with the Slum*, 353; C. B. J. Snyder, "Report of the Superintendent of School Buildings," in *Seventh Annual Report of the New York City Department of Education* (Records of the Department of Education, 1904), 286; *New York City Board of Education Annual Financial and Statistical Report, 1906–1908* (Records of the Department of Education, 1908).

51. Cooper talks about conflicts between climate-control and fresh-air-school advocates in her book on air-conditioning (see especially chapter 3). Mark Rose's book about the increased use of gas and electricity, *Cities of Light and Heat: Domesticating Gas and Electricity in Urban America* (University Park: Pennsylvania State University Press, 1995), is a notable exception to this limitation in the historical literature. Though his book is not primarily about schools, he does contextualize the pitch for increased utility use within concerns for the physical environment of educational institutions (see especially 99–106).

52. This literature is both sophisticated and extensive. Lawrence Cremin's *The Transformation of the School: Progressivism in American Education, 1876–1957* (New York: Alfred A. Knopf, 1961) was seminal. The book argued that a faith in science and management combined with a broad social reform agenda to give educators a new set of goals. Since Cremin's study, authors have struggled to determine the relevant actors who shaped Progressive education. I offer a small sampling of this scholarship here. For authors who portrayed Progressive educators as agents of social control, see David B. Tyack, *The One Best System: A History of American Urban Education* (Cambridge: Harvard University Press, 1974), and Dominick Cavallo, *Muscles and Morals: Organized Playgrounds and Urban Reform, 1880–1920* (Philadelphia: University of Pennsylvania Press, 1981). While Cavallo looks to distinguish himself from this school of thought, and does see a less nefarious reform movement, he writes about a playground system that sought to do critical socializing work. Herbert M. Kliebard's *The Struggle for the American Curriculum, 1893–1958* (New York: Routledge, 1995) does not fundamentally challenge this top-heavy, elite model, but he does disagree that there was a monolithic Progressive consensus. Rather, he sees the school as a hotly contested curricular battleground. Finally, revisionist scholarship has sought a voice for parents and students. See Stephen Hardy and Alan G. Ingham, "Games, Structures, and Agency: Historians on the American Play Movement," *Journal of Social History* 17 (1983): 285–301, and William J. Reese, *Power and the Promise of School Reform: Grassroots Movements during the Progressive Era* (Boston: Routledge and Kegan Paul, 1986).

53. Louise Dunham Goldsberry, "The Open-Air School and Out-Door Education" (1921), 64, 115, Prints and Photographs Division, Library of Congress.

54. Ibid., chapter 11; boxes 76, 89, 99, 117, Goldsberry Collection of Open-Air Schools, Prints and Photographs Division, Library of Congress.

55. Goldsberry, "The Open-Air School," 123, box 42, Goldsberry Collection of Open-Air Schools.

56. Boxes 12 and 45, folder 2, Goldsberry Collection of Open-Air Schools. See also boxes 11, 16, and 52.

57. Boxes 8, 18, 20, 32, 39, 40, and box 45, folder 1, Goldsberry Collection of Open-Air Schools.

58. David Rothman's *Conscience and Convenience: The Asylum and Its Alternatives in Progressive America* (Boston: Little, Brown and Co., 1980) gives an excellent account of the era's attempt to reform its prisons. Concerned that incarceration was ineffective and convinced that environment, not innate evil, led to illegal behavior, Progressives sought to remake their institutions in a way that could cure the criminal. This is an excellent corollary to the preventorium model (although in this case, the problem was far more physical than mental), which assumed that environment, not simply invading microbes, left children susceptible to illness and that a well-planned school could help them. For a history of the preventorium, see Cynthia A. Connolly, *Saving Sickly Children: The Tuberculosis Preventorium in American Life, 1909–1970* (New Brunswick: Rutgers University Press, 2008).

59. Boxes 59, 28, 72, 54, Goldsberry Collection of Open-Air Schools.

60. Boxes 48 and 69, box 45, folders 1 and 2, boxes 25B, 32, 61, Goldsberry Collection of Open-Air Schools.

61. Box 54, Goldsberry Collection of Open-Air Schools.

62. Boxes 20, 25A, 25B, 35, 72, Goldsberry Collection of Open-Air Schools.

63. Boxes 35, 25B, 31, 59, 58, 45, Goldsberry Collection of Open-Air Schools.

Chapter Two

1. National Carbon Company, "Sunshine Map," 1929, Geography and Map Division, Library of Congress.

2. For a solid account of modernist advertising, see Roland Marchand, *Advertising the American Dream: Making Way for Modernity, 1920–1940* (Berkeley: University of California Press, 1985). Chapter 8 examines light generally, and chapter 9 the sun and sunlight specifically. See also, Pamela Walker Laird, *Advertising Progress: American Business and the Rise of Consumer Marketing* (Baltimore: Johns Hopkins University Press, 1998), and Jackson Lears, *Fables of Abundance: A Cultural History of Advertising in America* (New York: Basic Books, 1994). Laird cautions against relying too heavily on advertisements in historical inquiry. She argues that pitches tell more about the values of advertisers than the culture in which they appeared. This reading of the evidence is most persuasive; here, I hope that the nonpromotional evidence I provide gives a context that allows me to avoid the pitfalls about which she writes.

3. Theobald A. Palm, "The Geographical Distribution and Aetiology of Rickets," *Practitioner* 45 (1890): 270–79, 321–42.

4. Edward Mellanby, "An Experimental Investigation on Rickets," *Lancet* 196 (1919): 407–12; Alfred Hess, G. F. McCann, and A. M. Pappenheimer, "Experimental Rickets in Rats: The Failure of Rats to Develop Rickets on a Diet Deficient in Vitamin D," *Journal of Biological Chemistry* 47 (1921): 395–409.

5. E. V. McCollum et al., "Studies on Experimental Rickets: An Experimental Demonstration

of the Existence of a Vitamin Which Promotes Calcium Deposition," *Journal of Biological Chemistry* 53 (1922): 293–312.

6. It turns out that vitamin D's primary effect is not in the bones. It is most important for its ability to help the intestines absorb calcium into the body. Calcium is spectacularly important, playing key roles in processes from heart-muscle contraction to neuronal signaling. In fact, bones weaken with calcium deficiencies because the body takes the critical mineral from its best repository, the bones, and redistributes it to places it is more pressingly needed. The resulting skeletal weakness is a necessary evil, considering the far more disastrous alternative.

7. Hess, McCann, and Pappenheimer, "Experimental Rickets in Rats." The skin is a fantastic vitamin D factory. Milk, now fortified with D, provides about 100 international units (IU) of D. The recommended daily allowance is 400 IU. Twenty minutes in a bathing suit at the beach with no sunscreen provides the body with 10,000–20,000 units.

8. Ibid.; McCollum et al., "Studies on Experimental Rickets."

9. As recent concerns about the ozone layer indicate, not all parts of the atmosphere are similarly opaque to all regions of the solar spectrum. Ozone is a particularly effective ultraviolet blocker. This was only vaguely understood in the 1920s, and the characterization here presented still holds. Sunlight from the horizon will have to pass through more of all parts of the atmosphere.

10. W. W. Coblentz, "Biologically Active Component of Ultra-Violet in Sunlight and Daylight," *Illuminating Engineering Society Transactions* 26 (1931): 572–77; Henry Laurens, "Factors Influencing the Choice of a Source of Radiant Energy," *Journal of the American Medical Association (JAMA)* 103 (1934): 1447–52; Matthew Luckiesh, "Ultraviolet in Sunlight," *Scientific American*, October 1922, 258.

11. Charles Sheard, "Ultraviolet Radiation and Its Transmission by Various Substances," *JAMA* 88 (1927): 1315–18; Arthur Knudson, "Rickets," *American Journal of the Diseases of Children* 44 (1932): 531–41. See also Frederick F. Tisdall and Alan Brown, "Relation of the Altitude of the Sun to Its Antirachitic Effect," *JAMA* 92 (1929): 860–64.

12. C. P. Obenschain, "Therapeutic Actions of Light," *Virginia Medical Monthly* 56 (1929): 59; Edgar Mayer, *Clinical Application of Sunlight and Artificial Radiation, Including Their Physiological and Experimental Aspects with Special Reference to Tuberculosis* (Baltimore: Williams and Wilkins Co., 1926), 182–83; H. E. Kleinschmidt, "Sun Baths in a Health Camp," *Nation's Health* 6 (1924): 691–92; Coblentz, "Biologically Active Component," 572–75; Frederick F. Tisdall, "Sunlight and Health," *American Journal of Public Health* 16 (1926): 694–99.

13. Caleb Williams Saleeby: *The Eugenic Prospect: National and Racial* (New York: Dodd, Meade, and Co., 1921), chapter 18, and *Sunlight and Health*, 5th ed. (New York: G. P. Putnam's Sons, 1924), 6–11. Historians of pollution have not treated this part of the smoke problem. As my upcoming arguments about sun substitutes indicate, the association of pollution with ill health did not necessarily require legislative intervention. Instead, many suggested, technological innovation could render smoke harmless. For solid histories of pollution, see Tarr, *Search for the Perfect Sink*, and Stradling, *Smokestacks and Progressives*.

14. Fred O. Tonney, Gerald L. Hoeft, and Paul P. Somers, "Loss of Actinic Intensity in Urban Sunshine due to Air Pollution," *Journal of Preventive Medicine* 4 (1930): 139–48; "Sunshine Lost at Noon," *JAMA* 93 (1929): 1230; J. H. Shrader, Maurice H. Coblentz, and Ferdinand A. Korff, "Effect of Atmospheric Pollution upon Incidence of Solar Ultra-Violet Light," *American Journal of Public Health* 19 (1929): 717–24; "New Tests Will Show How Soot Blots Out Sun," *Chicago Daily Tribune*, 4 May 1930, 26; "Machines Set up by U.S. to Test Chicago Smoke," *Chicago Daily*

Tribune, 1 March 1932, 13; "Meller Foresees Smoke Control by Air Hygiene Zones," *Washington Post,* 5 March 1933, 15; W. A. Evans's "How to Keep Well: Sunlight," *Washington Post,* 8 May 1925, 10, and his "How to Keep Well: What Britain Is Doing to Halt Air Pollution," *Washington Post,* 9 April 1927, 12; Dorothy Pletcher, "Are We to Have Sunless Cities," *Washington Post,* 15 December 1929, SM17; Jacques W. Redway, "Air Pollution," *New York Times,* 24 May 1929, 23; H. B. Meller, "The Sources of Air Pollution and the Methods of Regulating It," *New York Times,* 17 January 1931, XX7.

15. There were—and are—two main units for measuring the spectral qualities of light: the millimicron (mμ) and the angstrom (10 angstroms = 1 mμ).

16. Eugene H. Smith to J. Cecil Alter, 29 October 1924, box 2452, General Correspondence of the Weather Bureau, 1912–1942, Weather Bureau, National Archives at College Park, MD. See also from the same collection Mary E. Martin to Weather Bureau, 27 March 1927, box 2807; F. R. Edwards to H. H. Kimball, 1925; and W. D. Thurber to Weather Bureau, 29 May 1926, box 2808; as well as "Kegel Attacks City's Smoke Pall as Peril to Citizens' Health," *Chicago Daily Tribune,* 28 February 1929, 8; Pletcher, "Are We to Have Sunless Cities," SM17; "Says Smoke Pall Impedes Health Rays," *New York Times,* 8 October 1927, 2.

17. Weather Bureau to W. D. Thurber, 3 June 1926, "Memorandum," 24 June 1927, box 2807, General Correspondence of the Weather Bureau, 1912–1942; "Air Dust Census about to Begin," *New York Times,* 27 May 1928, 112.

18. C. F. Varbin to Charles J. McCabe, 16 January 1926, box 2808, General Correspondence of the Weather Bureau, 1912–1942; see also from that collection C. C. Clark to Ralph M. Shaw, 28 June 1933, and C. F. Marvin to Mrs. Lester G. Seacat, 13 March 1931, box 2997.

19. Caleb Williams Saleeby, "From Heliotherapy to Heliohygiene" (paper presented at the First International Conference on Light, Leysin, 10–13 September 1928), 509; Consolidated Gas Company, "Free New York from Smoke," *New York Times,* 4 January 1928, 26; Solvay Coke's "Consider Your Health," *Chicago Daily Tribune,* 25 March 1930, 12, and "Change Now . . . to Chicago Solvay Smoke," *Chicago Daily Tribune,* 12 March 1930, 13; Weil-McLain Boilers, "Is Smoke Robbing Our Children of Health-Giving Sunshine?," *Chicago Daily Tribune,* 8 October 1927, 12.

20. "Free Medicine Going to Waste," *Los Angeles Times,* 7 December 1925, A22; W. A. Evans, "How to Keep Well: Window Glass as Man's Enemy," *Washington Post,* 4 December 1926, 14; see also Evans's "How to Keep Well: Light Treatments," *Chicago Daily Tribune,* 26 March 1926, 8, and his "How to Keep Well: Light Treatments," *Washington Post,* 26 March 1926, 12. In addition, "Visitors Inspect 50 Hospitals Here," *New York Times,* 13 May 1927, 22; "Sun's Rays Have Healing Powers," *Los Angeles Times,* 25 November 1925, 17; "Sun Baths Urged as Best Physician to Sickly Children," *Washington Post,* 9 August 1925, 17; Barclay L. Severns, "Health in Sunshine," *Los Angeles Times,* 19 March 1923, III3.

21. Rickets folder, Science Service Collection, Division of Medicine and Science, National Museum of American History, Smithsonian Institution. For information about the rising tide of concern for the health of babies, see Richard A. Meckel, *Save the Babies: American Public Health Reform and the Prevention of Infant Mortality, 1850–1929* (Ann Arbor: University of Michigan Press, 1998).

22. According to best current medical practice, this racial disparity is correct. Other racial arguments, however, were of a more dubious sort; I will take a closer look at some of those in chapter 4.

23. Alfred F. Hess and Lester J. Unger, "Prophylactic Therapy for Rickets in a Negro Community," *JAMA* 69 (1917): 1583–86; Edwin T. Wyman and Charles A. Weymuller, "Organization of

a Special Clinic," *JAMA* 83 (1924): 1479–83; John Howland, "Starving for Sunshine," *Delineator*, March 1926, 16; "Babies and Sunshine," *Literary Digest*, 1 December 1923, 27; "Violet Rays to Help Crippled Children," *New York Times*, 15 October 1924, 25; Evans, "How to Keep Well: Light Treatments," *Chicago Daily Tribune*, 8; Royal S. Copeland, "How to Be Healthy: What You Can Do to Help a Baby Who Has 'Rickets,'" *Washington Post*, 26 October 1922, 8; Lulu Hunt Peters, "Diet and Health: Rickets and Sunshine," *Los Angeles Times*, 27 March 1925, A6.

24. Mayer, *Clinical Application*, 226–27; F. W. Parsons, "Sun Worship," *Saturday Evening Post*, 16 November 1929, 28.

25. Charles Greeley Abbot, *The Sun and the Welfare of Man*, vol. 2, Smithsonian Scientific Series (New York: Smithsonian Institution Series, 1929), 235.

26. See Downes and Blunt's "Researches on the Effect of Light upon Bacteria," 497, and their "On the Influence of Light upon Protoplasm," 199, 207–8. Paul Starr's *Social Transformation of American Medicine* (New York: Basic Books, 1982) most clearly articulates this standard historical position. See also Judith Walzer Leavitt, *Typhoid Mary: Captive to the Public's Health* (Boston: Beacon Press, 1996); Elizabeth Fee and Evelynn M. Hammonds, "Science, Politics, and the Art of Persuasion: Promoting the New Scientific Medicine in New York City," in *Hives of Sickness: Public Health and Epidemics in New York City*, ed. David Rosner (New Brunswick: University of Rutgers Press, 1995), 155–96. There are notable exceptions to the reductionism story. Other historians have argued that, in reality, doctors through the '30s often saw disease as a product of a weak constitution or sought treatments that healed the whole body. Often this argument contextualizes holistic doctors as people concerned with a troubling modernity. See David Cantor, ed., *Reinventing Hippocrates* (Aldershot: Ashgate Publishing Co., 2002); Christopher Lawrence and George Weisz, eds., *Greater than the Parts: Holism in Biomedicine, 1920–1950* (New York: Oxford University Press, 1998); David Cantor, "The Diseased Body," in *Medicine in the Twentieth Century*, ed. Roger Cooter and John Pickstone (Amsterdam: Harwood Academic Publishers, 2000), 347–66. Also of note, some studies of the history of vitamins have argued that bacteriology helped push deficiency diseases to the forefront of scientific research. In essence, doctors applied the new paradigm, in which specific agents caused specific conditions to nonbacterial ailments. See Christiane Sinding, "The History of Resistant Rickets: A Model for Understanding the Growth of Biomedical Knowledge," *Journal of the History of Biology* 22 (1989): 461–95; K. Codell Carter, "The Germ Theory, Beriberi, and the Deficiency Theory of Disease," *Medical History* 21 (1977): 119–36. For an overview of the vitamin craze, see Rima D. Apple, *Vitamania: Vitamins in American Culture* (New Brunswick: Rutgers University Press, 1996).

27. Sir Henry Gauvain, foreword to *Ultra-Violet Rays*, by Percy Hall (London: William Heinemann Ltd., 1930), vi. It was not uncommon for interwar doctors to turn to Hippocrates as a model. David Cantor's *Reinventing Hippocrates* provides ample scholarship on the resuscitation of the ancient doctor.

28. Notwithstanding arguments in favor of sunlight's ability to lift spirits and ease depression, its primary biophysical role today is in the production of vitamin D. Hess and his contemporaries were roughly right. Sunlight catalyzes the transformation of a precursor found in the skin into vitamin D (or to be more precise, one of the forms of D, as there are others). From there, the blood carries the molecule to the liver, and eventually the kidneys where it undergoes further chemical changes before it is deployed to the intestines where it regulates calcium intake.

29. See Frank H. Krusen's "Present Day Problems in Light Therapy," *New York State Journal of Medicine* 33 (1933): 1154–55, and his *Light Therapy* (New York: Paul Hoeber, 1933), chapters 2, 7.

30. Anne Frank, *The Diary of a Young Girl* (New York: Penguin Books, 2007), 318.

31. Profile on Niels Ryberg Finsen from the Nobel Prize e-museum, http://nobelprize.org/medicine/laureates/1903/press.html (accessed 20 June 2006) and http://nobelprize.org/medicine/laureates/1903/finsen-bio.html (accessed 20 June 2006); see also "How Diseases Are Treated in Europe," *New York Times*, 27 October 1901, 12; "Finsen's Healing Rays," *Chicago Daily Tribune*, 3 February 1903, 6.

32. Paul De Kruif, *Men against Death* (New York: Harcourt Brace and Co., 1932), chapter 10; Niels R. Finsen, *Phototherapy: (1) the Chemical Rays of Light and Small-Pox, (2) Light as a Stimulant, (3) the Treatment of Lupus Vulgaris by Concentrated Chemical Rays*, trans. James H. Sequeria (London: Arnold, 1901).

33. De Kruif, *Men against Death*, chapter 11; Paul De Kruif, "Old Doctor Sun," *Ladies' Home Journal*, October 1931, 6, 7.

34. De Kruif, *Men against Death*, 300. Bernhard did have other enthusiastic supporters; see Friedrich Ellinger, *The Biological Fundamentals of Radiation Therapy: A Textbook*, trans. Reuben Gross (New York: Elsevier Publishing Co., 1941); Walter H. Eddy, "Sunbrown versus Sunburn," *Good Housekeeping*, June 1934, 98; Edgar Mayer, "The Fundamentals and the Clinical Aspects of Light Treatment," *JAMA* 89 (1927): 361–67.

35. Benjamin Goldberg, "Heliotherapy," *Archives of Physical Therapy* 11 (1930): 263; R. I. Harris, "Heliotherapy in Surgical Tuberculosis," *American Journal of Public Health* 16 (1926): 687–99; A. Gottlieb, "Natural Heliotherapy," *Archives of Physical Therapy* 10 (1929): 110.

36. Later in this chapter, I will directly address the curious fact that these were employees of lighting companies celebrating sunlight. See Matthew B. Luckiesh and August John Pacini, *Light and Health: A Discussion of Light and Other Radiations in Relationship to Life and to Health* (Baltimore: Williams and Wilkins Co., 1926). See also Auguste Rollier and A. Rosselet, *Heliotherapy* (London: Henry Frowde and Hodder and Stoughton, 1923); Franz Thedering, *Sunlight as Healer: A Popular Treatise* (Slough, England: Sollux Publishing Co., 1926); Eleanor Hilda Russell and William Kerr Russell, *Ultra-Violet Radiation and Actinotherapy*, 3rd ed. (Edinburgh: E. and S. Livingstone, 1933); Krusen, *Light Therapy*; Robert Aitken, *Ultra-Violet Radiations and Their Uses* (Edinburgh: Oliver and Boyd, 1930).

37. George H. Maughan and D. F. Smiley, "Irradiations from a Quartz-Mercury-Vapor Lamp as a Factor in the Control of Common Colds," *American Journal of Hygiene* 9 (1929): 466–72; G. H. Maughan and Dean F. Smiley, "Effect of General Irradiation with Ultra-Violet Light upon the Frequency of Colds," *Journal of Preventive Medicine* 2 (1928): 69; "Curing a Cold," *Washington Post*, 14 February 1928, 6; W. A. Evans, "How to Keep Well: Preventing Colds by Ultraviolet Rays," *Chicago Daily Tribune*, 21 March 1928, 10; "Cornell to Fight Colds," *New York Times*, 3 November 1930, 30.

38. Thomas Victor Tanguy, *The Science and Practice of Dental Actinotherapy* (St. Louis: C. V. Mosby Co., 1928); F. Talbot, *Actinotherapy for Dental Diseases* (London: John Bale, Sons and Danielsson, Limited, 1928); August John Pacini, *Ultra-Violet Energy in Dentistry: Biophysical Studies* (Chicago: Victor X-Ray Corporation, 1923); W. H. Eddy, "Nature Gives Us Sunlight—Let's Use it!," *Good Housekeeping*, July 1938, 51; Arnold Lorand, *The Ultra-Violet Rays: Their Action on Internal and Nervous Diseases and Use in Preventing Loss of Color and Falling of the Hair* (Philadelphia: F. A. Davis Co., 1928); Hair and Scalp Institute, "The Progress of Science," *Chicago Daily Tribune*, 19 June 1928, 20. For a sampling of similar ads, see, from the Wellington Hair and Scalp Institute, "Your Hair . . . Watch It Grow," *Chicago Daily Tribune*, 6 December 1931, 24, "Ultra Violet Rays Grow Hair!," *Chicago Daily Tribune*, 17 April 1928, 24, and "Shabby Hair?,"

Chicago Daily Tribune, 6 December 1927, 24; from the Thomas [Hair and Scalp Corporation], "The Thomas' World's Greatest Scalp Specialists," *Chicago Daily Tribune*, 4 October 1927, 24, and "Thomas Gives Ultra-Violet Scalp Treatments," *Chicago Daily Tribune*, 25 October 1927, 8; and from the Hair and Scalp Institute, "Did You Hear this Over W-G-N Last Night?," *Chicago Daily Tribune*, 19 September 1928, 22.

39. Russell L. Cecil, ed., *A Text-Book of Medicine* (Philadelphia: W. B. Saunders Co., 1929); John H. Musser, ed., *Internal Medicine: Its Theory and Practice* (Philadelphia: Lea and Febiger, 1934); Heinrich F. Wolf et al., *Textbook of Physical Therapy* (New York: D. Appleton-Century Co., 1933); H. H. Conybeare, ed., *A Textbook of Medicine* (Edinburgh: E. and S. Livingstone, 1929); Howard Thomas Karsner and Simon Flexner, *Human Pathology: A Textbook* (Philadelphia: Lippincott, 1935); Kenneth L. Burdon, *A Textbook of Bacteriology* (New York: MacMillan Co., 1932). Scientific articles also illustrate this divisiveness; see Mayer, "The Fundamentals and the Clinical Aspects of Light Treatment"; Frank H. Krusen, "Medical Application of Ultra-Violet Radiant Energy," *Annals of Internal Medicine* 14 (1940): 641–51; "Sunlight and Anemia," *Archives of Physical Therapy* 12 (1931): 427–31; Frank Albert Davis, "Physiotherapy in the Treatment of Tuberculosis with Report on Certain Laboratory Findings Relating to the Physiological Action of the Ultra Violet Rays on Blood Chemistry," *United States Veterans' Bureau Medical Bulletin* 1 (1925): 22–29.

40. Shortwave ultraviolet light is almost completely filtered by conventional glass, and that is what these scientists were concerned about. Today, of course, we think more about how sunlight can fade fabrics or paintings. The longest waves of the ultraviolet part of the spectrum are responsible for this fading (conventional windows block shorter waves). In the 1950s, museums and stores began to focus attention on this problem, and companies began selling special clear coatings that they said would block out waves harmful to displays and masterpieces. William M. Freeman, "Light Absorbers Now on Market," *New York Times*, 9 November 1957, 26; Meyer Berger, "About New York," *New York Times*, 10 August 1956, 18.

41. Charles Sheard, "Cheeks of Tan," *Hygeia* 5 (1927): 221.

42. Vita Glass, "Do You Still Live in a Cave?," *New York Times*, 7 October 1928, SM15.

43. See the Vita Glass ads "Prisoners of Glass," *New York Times*, 16 September 1928, 86, and "You Spend Your Day among Thieves," *Chicago Daily Tribune*, 25 November 1928, B2.

44. See the Vita Glass ads "Where Did you Get That Tan . . . ?," *New York Times*, 11 March 1929, 17, and "Do You Believe in *Sunlight*?," *New York Times*, 2 December 1929, 21.

45. The Quartz-Lite advertisements include "Real Sunlight for Work Rooms," *Chicago Daily Tribune*, 27 October 1927, 18; "Nature's Invisible Tonic for 4 O'Clock Fatigue," *Chicago Daily Tribune*, 10 November 1927, 12; "Make Every Room a Sun Room," *New York Times*, 8 September 1927, 18; and "—The Low Price," *Chicago Daily Tribune*, 28 July 1927, 11.

46. See the Lustra-glass sales pamphlets *Specification Sheet for 1930*, *Lustraglass Looks Like Plate Glass, Sells at Window Glass Prices*, and *Lustra-glass: The Ultra-Violet Ray Glass*, Collected Sales Catalogs, Trade Literature, National Museum of American History Library, Smithsonian Institution.

47. Albert G. Ingalls, "Ultra-Violet Transmitting Glass—Has It Made Good?," *Scientific American*, April 1929, 338; "Report on Window Glass Substitutes," *JAMA* 88 (1927): 1562–67. Nancy Tomes, "Merchants of Health: Medicine and Consumer Culture in the United States, 1900–1940," *Journal of American History* 88 (2001): 519–47, gives a great account of the intersection of professional medicine and consumerism.

48. W. W. Coblentz and R. Stair, "The Effect of Solarization upon the Ultraviolet Transmis-

sion of Window Materials," *Illuminating Engineering Society Transactions* 23 (1928): 1121–51; George W. Caldwell, "The Clinical Value of Sunlight through Ultraviolet Transmitting Glass," *JAMA* 92 (1929): 2088–90. Solarization created major problems for strictly prescribed light treatments, an issue that I will treat in chapter 3.

49. Ingalls, "Ultra-Violet Transmitting Glass," 388; W. W. Coblentz, "Ultraviolet Transmitting Glasses: Specification of Minimum Transmission," *JAMA* 95 (1930): 864–65.

50. Alpine Sun Lamp, "Safe, Simple and Effective," *Good Housekeeping*, December 1929, 241.

51. Sunbeam, "Give Your Children Strength and Brimming Vitality," *Chicago Daily Tribune*, 9 March 1930, 18; Eveready, "Keep Your Summer Health and Tan," *Chicago Daily Tribune*, 8 December 1929, D7; General Electric, "Give Them Summer Sunlight All Winter Long," *Chicago Daily Tribune*, 13 December 1933, 18; "'Pal, Come Get Your Sun Bath, Too!'" *New York Times*, 11 January 1933, 8, "The Sun's Gone on His Winter Vacation," *New York Times*, 25 January 1933, 8; Commonwealth Edison Electric Shops, "Florida Sunshine at Home," *Chicago Daily Tribune*, 8 February 1933, 6; Commonwealth Edison Electric Shops, "Sun Starvation," *Chicago Daily Tribune*, 20 November 1930, 20; General Electric, "That's Why Chicago Homes Need Indoor Sunlight," *Chicago Daily Tribune*, 15 December 1933, 10; General Electric, "You Can Have a Summer Seaside Tan Right Now at Home!," *New York Times*, 16 May 1930, X15.

52. "Even Nudists Can't Get Enough Sunshine in Winter without a Sun Lamp," *New York Times*, 24 January 1934, 14; "The Winter Sun Is an Old Swindler," *New York Times*, 31 January 1934, 12; Eveready, "Growing Children Need Sunshine in Winter Too," *Hygeia* 9 (1931): 9; General Electric, "You Can Have a Summer Seaside Tan," X15; Eveready, "Why Risk Your Health in Winter," *New York Times*, 10 November 1929, RP36; and "Why Do We Dread the Cold, Drear[y] Days of Winter," *Chicago Daily Tribune*, 17 November 1929, E9.

53. Health Ray, "Is the Sun in Your Home?," *New York Times*, 4 January 1932, 44; L and H Super-Sun, "Man Has Created a Sun," *New York Times*, 13 May 1928, 83; Super-Sun, "Absorb Glorious Vitality from the Man Created Sun," *New York Times*, 10 November 1929, SM12.

54. National Carbon Company, "Abstracts—Ultra-Violet and Other Radiation from Carbon Arc," binder of collected materials housed at National Library of Medicine, National Institutes of Health. The version I saw had abstracts from 1942, but the first assembly of materials probably occurred in the late '30s.

55. Wolfgang Schivelbusch, *Disenchanted Night: The Industrialization of Light in the Nineteenth Century* (Berkeley: University of California Press, 1988); David E. Nye's *American Technological Sublime* (Cambridge: MIT Press, 1994), and his *Electrifying America: Social Meanings of a New Technology, 1880–1940* (Cambridge: MIT Press, 1990); John A. Jakle, *City Lights: Illuminating the American Night* (Baltimore: Johns Hopkins University Press, 2001); A. Roger Ekirch, *At Days Close: Night in Times Past* (New York: W. W. Norton and Co., 2005), see especially the epilogue. Among these titles, Nye's *Electrifying America* is unique for its recognition that Americans valued natural light and considered their ability to make a superior surrogate for nature (see especially chapter 8). Matthew Luckiesh figures prominently in Nye's text.

56. General Electric, "Don't Sun-Starve Your Baby!," *Good Housekeeping*, January 1932, 96; General Electric, "You Can Have a Summer Seaside Tan," X15; Commonwealth Edison Electric Shops, "Sun Starvation," 20; Vita Glass, "Is This Little Fellow Worth $25?," *New York Times*, 5 February 1929, 19.

57. A. H. Pfund, "A Practical Window for Transmitting Ultraviolet Rays," *JAMA* 91 (1928): 18–19.

58. "Lamp Sales in 1932," *Edison Electric Institute Bulletin* 1 (1933): 145–47; "Incandescent

Lamp Statistics, 1933," *Edison Electric Institute Bulletin* 1 (1934): 335, 352. Special thanks to
the General Electric Archives at the Schenectady Museum and Suits-Bueche Planetarium. Tony
Scalise helped me navigate the materials and Chris Hunter helped me compute these figures.

59. John Sadar, "The Healthful Ambience of Vitaglass: Light, Glass and the Curative Environment," *Architectural Research Quarterly* 12, nos. 3/4 (2008): 269, 281.

60. Ibid.

61. Vita Glass, "Why Have a Window in Your Office," *Wall Street Journal*, 14 May 1929, 7. I
must recognize that Nash is primarily writing about wilderness not nature, but his analysis ap-
plies here nonetheless. Roderick Nash, *Wilderness and the American Mind* (New Haven: Yale University Press, 1982). In his article "The Trouble with Wilderness," William Cronon approaches
this argument. He contends that wilderness, as a concept, is troubling because it is impossible to
create two oppositional categories, natural and unnatural. Any attempt to do so eliminates the
middle ground upon which people may actually build a livable world. Cronon's essay began a
period in which hosts of historians reconsidered the usefulness of the concept nature. Matthew
Gandy's *Concrete and Clay: Reworking Nature in New York City* (Cambridge: MIT Press, 2002)
brings this sort of thinking to a study of New York and its surrounding area. For Gandy, natural
and urban are not oppositional; rather, Gotham and its wilderness are interconnected. William
Cronon, "The Trouble with Wilderness; or, Getting Back to the Wrong Nature," in *Uncommon
Ground: The Human Place in Nature*, ed. William Cronon (New York City: W. W. Norton and
Co., 1996), 69–90.

62. "The UVIARC Sun Lamp for Healthful Sun Baths," 1929; "Instruction for Setting Up and
Operation of UVIARC Sun Lamp on Alternating Current," General Electric Vapor Lamp Co.,
Trade Literature, National Museum of American History, Smithsonian Institution; Eveready,
"Growing Children Need Sunshine in Winter Too," 9.

63. Marchand, *Advertising the American Dream*, see especially 223–28.

64. "The UVIARC Sun Lamp for Healthful Sun Baths."

Chapter Three

1. Ernest Hemingway, *A Farewell to Arms* (London: Arrow Books, 2004), 89, 106.

2. Morris Fishbein, *The Medical Follies* (New York: Boni and Liveright, 1925).

3. Clarence A. Smith, "Uses and Abuses of Ultraviolet Rays," *Northwest Medicine* 28 (1929):
261–63; J. Harry Bendes, "Heliotherapy," *Minnesota Medicine* 5 (1922): 302–7; Miland E.
Knapp, "Ultra-Violet Therapy," *Minnesota Medicine* 22 (1939): 394–96; R. H. Walker, "Treatment of Intestinal Tuberculosis," *West Virginia Medical Journal* 32 (1936): 250–51 (paper origi-
nally present to the Kanawha Medical Society); Obenschain, "Therapeutic Actions of Light,"
58–60.

4. Ellinger, *The Biological Fundamentals of Radiation Therapy*; Mayer, *Clinical Application*;
Carleton Ellis and Alfred A. Wells, *The Chemical Action of Ultraviolet Rays* (New York: Chemi-
cal Catalog Co., 1925). For book-length treatises, see the following small sampling. In the com-
ing pages, I will reference particular aspects of practice, generally using titles not in this footnote,
but these works often offer similar pictures. Aitken, *Ultra-Violet Radiations and Their Uses*; Hall,
Ultra-Violet Rays; Francis Howard Humphris, *Artificial Sunlight and Its Therapeutic Uses*, 5th ed.
(London: Oxford University Press, 1929); Krusen, *Light Therapy*; Thomas Clyde McKenzie and
Alfred Alexander King, *Practical Ultra-Violet Light Therapy: A Handbook for the Use of Medical*

Practitioners (New York: William Wood and Co., 1926); Rollier and Rosselet, *Heliotherapy*; Russell and Russell, *Ultra-Violet Radiation and Actinotherapy*.

5. Edgar Mayer, "The Relation of Light to Health and Disease," *Archives of Physical Therapy, X-Ray, Radium* 13 (1932): 337; Julius William Sturmer, *The Modern Sun Cult* (Washington, DC: Smithsonian Institute, 1931), 191–206; Henri Laurens, *The Physiological Effects of Radiant Energy* (New York: Chemical Catalog Co., 1933), chapter 3.

6. Edwin T. Wyman et al., "Value of Different Types of Glass for Transmitting Ultraviolet Light," *American Journal of the Diseases of Children* 37 (1929): 473–82.

7. National Carbon Company, "Abstracts—Ultra-Violet and Other Radiation."

8. H. N. Bundesen, "Sunshine and Health," *Ladies' Home Journal*, August 1938, 48; Krusen, "Present Day Problems in Light Therapy," 1154–57; Percival Nicholson, "Status and Technic of Natural and Artificial Heliotherapy," *Archives of Pediatrics* 46 (1929): 406.

9. Clayton H. Sharp and William F. Little, "The Problem of the Definition and Measurement of the Useful Radiation of Ultraviolet Lamps," *Illuminating Engineering Society Transactions* 26 (1931): 728; Samuel H. Watson, "The Use and Abuse of Heliotherapy in Tuberculosis," *JAMA* 87 (1926): 1026–31; A. H. Taylor and L. L. Holladay, "Measurement of Biologically Important Ultra-Violet Radiation," *Illuminating Engineering Society Transactions* 26 (1931): 711–26, 737–43.

10. R. B. Bourdillon, "Notes on the Measurement of Ultra-Violet Radiation," *American Journal of Physical Therapy* 5 (1929): 515–18; Taylor and Holladay, "Measurement of Biologically Important Ultra-Violet Radiation."

11. Taylor and Holladay, "Measurement of Biologically Important Ultra-Violet Radiation"; M. Luckiesh and L. L. Holladay, "Fundamental Units and Terms for Biologically Effective Radiation," *Optical Society of America Journal* 23 (1933): 197–205; W. W. Coblentz, "Ultraviolet Radiation Useful for Therapeutic Purposes," *JAMA* 98 (1932): 1082–85.

12. In rare cases, doctors would test their own photosensitivity and extrapolate the results to their patients. A. H. Taylor, "Measurement of Erythemal Ultraviolet Radiation," *Optical Society of America Journal* 23 (1933): 61; C. M. Sampson, *Physiotherapy Technic: A Manual of Applied Physics* (St. Louis: C. V. Mosby Co., 1923); 185–86, McKenzie and King, *Practical Ultra-Violet Light Therapy*, 52; Krusen, *Light Therapy*, 73; Humphris, *Artificial Sunlight and Its Therapeutic Uses*, 113.

13. Rollier and Rosselet, *Heliotherapy*, chapter 3; Goldberg, "Heliotherapy"; Nicholson, "Status and Technic of Natural and Artificial Heliotherapy"; Kleinschmidt, "Sun Baths in a Health Camp."

14. This procedure bears some similarity to present-day radiation treatment for cancer. Both attempt to check cell growth. Radiation treatments, however, use different, more penetrating wavelengths, and they do greater damage to the mutated cells, preventing their too-fast growth, while healthier cells are better able to repair themselves after treatment.

15. Knapp, "Ultra-Violet Therapy"; Sampson, *Physiotherapy Technic*, chapter 10; Humphris, *Artificial Sunlight and Its Therapeutic Uses*, see especially chapters 3 and 4.

16. Mitman, "Geographies of Hope." Histories of the emergence of the modern hospital have tended to focus on the shift from smaller, community-based institutions that offered more haphazard, though in some respects far more personal care, to modern, scientific, bureaucratic facilities. See David Rosner, *A Once Charitable Enterprise: Hospitals and Healthcare in Brooklyn and New York, 1885–1915* (New York City: Cambridge University Press, 1982); Charles Rosenberg, *The Care of Strangers: The Rise of America's Hospital System* (New York City: Basic Books, 1987). My focus here is far more on the forms of treatment available in these institutions. Jeanne Susan

Kisacky does write about the environmental conditions predominating in the hospital, but the focus is primarily on air (light's clearest appearance is near the end of the work, where Kisacky discusses how hospitals stopped trying to replicate the natural environment). Jeanne Susan Kisacky, "An Architecture of Light and Air: Theories of Hygiene and the Building of the New York Hospital, 1771–1932" (PhD diss., Cornell University, 2000).

17. Rickets folder, Science Service Collection, Division of Medicine and Science, National Museum of American History, Smithsonian Institution.

18. Robert E. Fitzgerald, "Perrysburg—a Mecca for Heliotherapy," *American Journal of Public Health* 16 (1926): 893–95; W. A. Evans, "How to Keep Well: Sun's Rays are Still Better than Man's," *Chicago Daily Tribune*, 28 March 1928, 10; De Kruif, "Old Doctor Sun," 6–7.

19. Folder 116, Records of the Children's Bureau, Still Pictures, National Archives at College Park, MD; "Boy Scouts," *New York Times*, 18 March 1923, X10; "Say Sun Treatment Cures Tuberculosis," *New York Times*, 29 September 1922, 17. For general studies of tuberculosis and its treatment, see Katherine Ott, *Fevered Lives: Tuberculosis in American Culture since 1870* (Cambridge: Harvard University Press, 1996), and Rothman, *Living in the Shadow of Death*. Greg Mitman's *Breathing Space: How Allergies Shape Our Lives and Landscapes* (New Haven: Yale University Press, 2007) offers a nice account of sanitaria and the outdoor experience.

20. Folder 106, Records of the Public Health Service, Still Pictures, National Archives at College Park, MD; folder 116, Records of the Children's Bureau, Still Pictures; G. O. Basset, "Heliotherapy in Surgical Tuberculosis," *United States Veterans' Bureau Medical Bulletin* 3 (1927): 342–46; Mabel C. Ryan, "Section of Physiotherapy and Occupational Therapy," *United States Veterans' Bureau Medical Bulletin* 6 (1930): 73; Eleanor Fisher, "Heliotherapy in the Treatment of Laryngeal Tuberculosis," *United States Veterans' Bureau Medical Bulletin* 3 (1927): 1066–68.

21. Council on Medical Education and Hospitals (American Medical Association), "Survey of Tuberculosis Hospitals and Sanatoriums in the United States," *JAMA* 105 (1935): 1855–1916. In 1926, the Tuberculosis Sanatorium Conference of Metropolitan New York undertook a similar statewide survey and found that, of thirty-two institutions, only six gave neither natural heliotherapy nor artificial light treatments, while two provided the former, nine the latter, and fifteen both. See *New York City's Institutions for the Tuberculous: Clinics, Hospitals, Sanatoria, Preventoria, Day Camps and other Agencies* (New York: The Tuberculosis Sanatorium Conference of Metropolitan New York, 1926).

22. *Annual Medical Report of the Trudeau Sanitarium, 1922–1931*, National Library of Medicine, National Institutes of Health.

23. South Mountain was also sometimes known as Mont Alto in its early years, but for simplicity's sake, I have used the former name throughout my text. *Spunk* reported summary information in the beginning of each issue, including the number of patients and a general outline of facilities. For narratives of the institution's growth, see Kathryn Elizabeth Yelinek, *The History of South Mountain Restoration Center, 1901–2001* (Pennsylvania Historical and Museum Commission for the Pennsylvania Department of Public Welfare, 2001); "A Review of South Mountain," *Spunk*, March 1933, 17–18; "A Review of South Mountain State Sanatorium," *Spunk*, February 1933, 19. Yelinek and *Spunk* do not agree on all details. I took the summary numbers at the beginning of each issue as most reliable then pieced together the most plausible account of South Mountain's growth from these sources, relying mostly on their points of accord and avoiding matters about which they disagreed. The institution remained a sanatorium until the 1960s when the state legislature made South Mountain one of two new state geriatric centers.

24. "Outdoor Life," editorial, *Spunk*, May 1922, 23; Walter C. Klotz, "Rest and Exercise,"

Spunk, August-September 1921, 13–15; Charles F. Arroyo's "How to Take the Rest Cure," *Spunk*, September 1920, 6, and his "The Return to Activity," *Spunk*, October 1920, 10.

25. Allen K. Krause, "The Treatment of Tuberculosis," *Spunk*, January 1923, 17–20; W. G. Turnbull, "The Question Box," *Spunk*, August 1922, 22.

26. "Heliotherapy," *Spunk*, April 1925, 20–22; Joseph Walsh, "The Onset of Tuberculosis," *Spunk*, August 1923, 20–22; "Sunshine and Shadows," *Spunk*, September 1923, 19; "The Sun Cure," *Spunk*, September 1923, 27.

27. "Heliotherapy," *Spunk*, 20–22. This and the previous block quote demonstrate other aspects of sunlight thinking; most notably, its class and primitivist elements. Those will be treated more fully in the next two chapters.

28. Elizabeth Ramsden, "Hamburg's History," *Spunk*, October 1924, 16–17; W. L. Latimer, "The Children's Hospital," *Spunk*, November 1924, 20; B. S. Herben, "Mending Done by the Sun," *Spunk*, June 1924, 19–20; Henry A. Gorman, "What They Should Expect," *Spunk*, October 1924, 18–19; Elysie L. Cole, "Ourselves as Others See Us," *Spunk*, April 1924, 20.

29. Arshag Kerem, "To Thee O Sun-God," *Spunk*, November 1925; G. F. Favis, "Heliotherapy in Tuberculosis," *Spunk*, March 1926, 15; A. Craig, "Leysin," *Spunk*, February 1926, 15–18; "Hail, Helios," *Spunk*, June 1928, 16; "Beneficial Effects of Sun Bathing," *Spunk*, July 1927, 23; "Sunlight as a Disinfectant," *Spunk*, September 1927, 18; "The Ultra-violet Ray a Stimulus to the Defense Mechanism of the Body," *Spunk*, November 1927, 16.

30. H. B. Pirkle, "Sun Worship and Heliotherapy," *Spunk*, July 1929, 17–20, "Ole Doc Sunshine," editorial, *Spunk*, May 1929, 15–16; Katharine F. Blake, "Speaking of Sunlight," *Spunk*, April 1929, 25–26.

31. Robert A. Milliken, "Bone and Joint Tuberculosis," *Spunk*, January 1934, 24–26; S. F. Strain, "Tuberculosis of the Larynx," *Spunk*, February 1932, 20–22; "Unsupervised Sun Baths," *Spunk*, June 1933, 16; "Sun and Air," *Spunk*, May 1932, 16.

32. *Tuberculosis Sanitorium and Preventorium*, Harry A. Wilmer and Lois Parker, film recording, National Library of Medicine, National Institutes of Health.

33. Ibid.

34. Hilbert F. Day, "Sunshine Camp in Cambridge," *Hygeia* 5 (1927): 248–51; Kleinschmidt, "Sun Baths in a Health Camp"; Latimer, "The Children's Hospital," 20.

35. See the following Vita Glass ads: "Now Every Window May Be an Avenue to Health," *New York Times*, 12 April 1927, 18; "Now Every Window May Be an Avenue to Health," *Washington Post*, 21 April 1927, 4; "At Last—the Health of the Sun *through Your Windows!*," *New York Times*, 17 April 1927, SM14; "At Last—the Health of the Sun *through Your Windows!*," *Washington Post*, 28 April 1927, 4; "Your Office Windows: A Great Discovery Now Makes Them an Aid to Better Health," *New York Times*, 12 December 1927, 16; "The Offices of the Equitable Company of New York Will Receive the Health of Whole Sunlight through Windows of Vita Glass," *New York Times*, 27 July 1927, 17.

36. Tom Pettey, "Spring Style for Factory Drops Windows," *Chicago Daily Tribune*, 1 April 1931, 23; Frank McCoy, "Health and Diet Advice: Give Babies Sun Baths in Summer," *Los Angeles Times*, 5 July 1930, A5; "Factory Building with No Windows," *New York Times*, 14 December 1930, RE4. For an account of innovative architecture in workplaces, see Betsy H. Bradley, *The Works: The Industrial Architecture of the United States* (New York City: Oxford University Press, 1999).

37. Sanford DeHart, "Electrical Therapeutics in Industry," *Electrical World* 99 (1932): 816–17.

38. William R. Fogg, "Daylight Illumination of Industrial Buildings," *Architectural Forum* 51 (1929): 410; "Coat of Tan Is Ready for Congress," *Washington Post*, 14 February 1929, 8; "'Sun Lamps' To Tan Cheeks of Members of Congress," *New York Times*, 13 February 1929, 1; W. C. Kalb, "Carbon Arc Sunshine in Industry," *Engineering and Mining Journal* 130 (1930): 579–80; S. A. Easton, "All Year Sunshine for Mine Workers," *Mining and Metallurgy* 10 (1929): 564–65; "Substitute Sunshine for Miners," *Literary Digest*, 25 January 1930, 564–65.

39. Matthew B. Luckiesh, *Artificial Sunlight: Combining Radiation for Health with Light for Vision* (New York: Van Nostrand Co., 1930), chapter 1, 28–29.

40. Matthew Luckiesh, "Linking Light with Health," *Illuminating Engineering Society Transactions* 27 (1932): 1; Matthew B. Luckiesh, *Light and Work: A Discussion of Quality and Quantity of Light in Relation to Effective Vision and Efficient Work* (New York: Van Nostrand Co., 1924), 1–4; Luckiesh, "A Benefactor of Civilization," *Illuminating Engineering Society Transactions* 25 (1930): 323–24; Luckiesh, *Artificial Sunlight*, 1, 7, 12.

41. See the following GE pamphlets, from General Electric, Trade Literature, National Museum of American History, Smithsonian Institution: *New Knowledge of Seeing in the Classroom, Notes and Designs for General Electric Lighting Sales Institute Course: Home Lighting Course, Seeing Begins, It's Only the Beginning, New Knowledge of Seeing, Notes and Designs*; see also Matthew Luckiesh and Frank K. Moss, "The Applied Science of Seeing," *Illuminating Engineering Society Transactions* 28 (1933): 842–65; Luckiesh, *Light and Work*; Luckiesh, "It's Only the Beginning," *Magazine of Light*, April 1935, 9–17.

42. A. H. Taylor, "The Color of Daylight," *Illuminating Engineering Society Transactions* 25 (1930): 154–71; Luckiesh and Moss, "The Applied Science of Seeing," 843; Luckiesh, "It's Only the Beginning," 10.

43. *Official Guide Book of the World's Fair of 1934* (Chicago: Cuneo Press, 1934), see especially 27, 103, 129; Philip Kinsley, "Human Miracle of Light Enters New Era at Fair," *Chicago Daily Tribune*, 15 April 1933, 5; Louise Bargelt, "House of Tomorrow at Century of Progress Has Walls of Glass," *Chicago Daily Tribune*, 11 June 1933; Cheryl R. Ganz, *The 1933 Chicago World's Fair* (Urbana: University of Illinois Press, 2008); Robert W. Rydell, "The Fan Dance of Science: American World's Fairs in the Great Depression," *Isis*, 76 (1985): 525–42. There were four other major windowless buildings; see "Five Buildings at Exposition Are Windowless," *Chicago Daily Tribune*, 27 May 1933, 3.

44. W. R. Flounders, "Synthetic Daylight Aids Poor-Vision Pupils," *Electrical World* 103 (1934): 581.

45. Folder 121, Records of the Public Health Service, Still Pictures; Frank H. Broome, "What! A Windowless School," *Magazine of Light*, December 1937, 31–32.

46. See the Vita Glass ads "Vita-Glass Brings Health through Your Windows," *Washington Post*, 7 April 1927, 11, and "Vita-Glass Brings Health through Your Windows," *New York Times*, 10 March 1927, 19; also Walter H. Eddy, "The Use of Ultra-Violet Light Transmitting Windows," *American Journal of Public Health* 18 (1928): 1470–79, and Vita Glass, "The Offices of the Equitable Company," 17.

47. Lustra-glass, *Specification Sheet for 1930*.

48. "Urban School to Open Soon," *Los Angeles Times*, 30 August 1931, C5; "Urban for Boys," *Los Angeles Times*, 30 August 1931, C3; "Urban Academy Keeps Pace," *Los Angeles Times*, 31 January 1932, B13.

49. I will talk a bit more about the Race Betterment Conference's curious flirtation with sunlight theory in the next chapter.

50. John W. M. Bunker, "Light and Life" (paper presented at the Third Race Betterment Conference, Battle Creek, Michigan, 2–6 January 1928), 609–20; R. W. Butts, "Sandusky Subscribes to Better Lighting for Its School System," *Magazine of Light*, February 1937, 6–9; C. H. Gleason, "Open-Air School Pupils Benefit from Natural and Artificial Sunlight," *American City*, April 1937, 107; Philip S. Potter, "The Sunshine School," *Hygeia* 6 (1928): 568–70.

51. Benjamin Goldberg, "An Apparatus for the Mass Treatment of Children with Ultraviolet Rays," *JAMA* 93 (1929): 1377–78; Sally Joy Brown, "Crippled Children Taught to Forget Their Handicaps: How It Is Done at Chicago School," *Chicago Daily Tribune*, 1 September 1935, D10.

52. For more about vitamin-fortified milk, see chapter 5.

53. Sunbaths folder, Science Service Collection, Division of Medicine and Science, National Museum of American History, Smithsonian Institution.

54. Vita Glass, "These Lions in the London Zoo Chose Vita Glass by Instinct," *New York Times*, 5 May 1927, 13.

55. "Phototherapy at the Zoological Gardens," *JAMA* 90 (1928): 1580; J. S. Hughes and R. L. Pyscha, "A Preliminary Report of the Measurement of Variation of Energy in the 'Vita Spectrum' of the Sunshine in Kansas," *Illuminating Engineering Society Transactions* 23 (1928): 243; "Zoo Men Relieved as N'Gi Improves," *Washington Post*, 20 February 1932, 3; "A Cure for 'Cage Paralysis,'" *New York Times*, 11 September 1927, XX15; "Artificial Sunlight for a Zoo," *Literary Digest*, 2 January 1932, 28; "Healthy Ape 'Patient' Demonstrates Cure," *New York Times*, 14 June 1932, 43.

56. Irving J. Kauder to [Semon] Bache and Co., 1932, *Glass: The Ultra Violet—Vitamin D*, Parke-Davis, Trade Literature, National Museum of American History Library, Smithsonian Institution.

57. Frank Ridgway, "Farm and Garden: Ultra-Violet Ray's Good Effect on Chick and Egg," *Chicago Daily Tribune*, 30 December 1924, 12; Ridgway, "Farm and Garden: And Now Science Invents Sun Rays for Young Chicks," *Chicago Daily Tribune*, 28 March 1926, A12; W. A. Evans, "How to Keep Well: Glassless Windows," *Chicago Daily Tribune*, 14 October 1925, 8; "Making a Home for Hens," *Los Angeles Times*, 3 November 1929, G16; Ross H. Gast, "Little Farm Homes," *Los Angeles Times*, 11 May 1930, I13; *Glass: The Ultra Violet—Vitamin D; More Money in Eggs and Chicks: An Interesting and Non-Technical Discussion of Ultra-Violet Light and Its Effect on Winter Egg Production—and Growing Chicks*, Hanovia Chemical Manufacturing Company, Trade Literature, National Museum of American History, Smithsonian Institution; Lawrence C. Porter and J. P. Ditchman, *Ultraviolet for Poultry*, General Electric, Trade Literature, National Museum of American History, Smithsonian Institution.

58. "Sunshine and Skyshine," *JAMA* 86 (1926): 872–73; Herman Goodman, "Macrocosm and Microcosm," *Medical Journal and Record* 135 (1932): 329–30; M. J. Dorcas, "Ultraviolet in Industry," *Illuminating Engineering Society Transactions* 28 (1933): 575–89.

59. Hall, *Ultra-Violet Rays*. Hall's popular book enjoyed multiple printings in America and England.

60. "Statement on Ultraviolet Therapy," in *Handbook of Physical Therapy*, 2nd ed. (Chicago: AMA Press, 1936), 197; Ronald Millar, E. E. Free, and Edward Elway, *Sunrays and Health* (New York: R. M. McBride and Co., 1929), 120–21.

61. Harvey W. Wiley, "The Sun in Service," *Good Housekeeping*, November 1927, 217; H. Harris Perlman, "Mercury Vapor Quartz Light Therapy in Pediatrics," *Archives of Physical Therapy* 10 (1929): 365–69; Smith, "Uses and Abuses of Ultraviolet Rays," 262–63; O. Leonard Huddleston, "Use of Physical Therapy in Internal Medicine," *Burdick Syllabus* 19 (1943): 1–2.

62. "Statement on Ultraviolet Therapy"; H. E. Mock, introduction to *Handbook on Physical Therapy*, 7–10.

63. "Regulations to Govern Advertising of Ultraviolet Generators to the Public Only," *JAMA* 98 (1932): 400–401.

64. Harold M. F. Behneman, "Use of 'Cold Quartz' Light in General Practice," *Archives of Physical Therapy* 14 (1933): 72–78; "Acceptance of Sunlamps," *JAMA* 100 (1933): 1863–64. For examples of sunlamp acceptances and rejections, see "Sirian Ultraviolet Lamp Not Acceptable," *JAMA* 100 (1933): 338; "Cold-Quartz Generator Acceptable," *JAMA* 100 (1933): 573–74; "Burdick Super-Standard Air-Cooled Lamp Acceptable," *JAMA* 99 (1932): 388; "Super Alpine Sun Lamp Acceptable," *JAMA* 99 (1932): 389.

65. Matthew Luckiesh, "An Epochal Light Source," *Electrical World*, 26 October 1929, 835; "G.E. Mazda Sunlight Lamps, Types S-1 and S-2 Acceptable," *JAMA* 99 (1932): 32; General Electric, "The General Electric Sunlight (Type S-1) Lamp," *Hygeia* 8 (1930): 886; A. B. Oday and L. C. Porter, "The Use of Ultraviolet Sources for the General Illumination of Interiors," *Illuminating Engineering Society Transactions* 28 (1933): 121–52.

66. Breuer, "Sunlight," editorial, *Hygeia* 8 (1930): 704.

67. Krusen, "Present Day Problems in Light Therapy," 1170; Henry Laurens, "Factors Influencing the Choice of a Source of Radiant Energy: Ultraviolet, Luminous and Infra-Red," in *Handbook of Physical Therapy*, 203; see also Laurens, "Factors Influencing the Choice of a Source of Radiant Energy," 1447–52.

68. Fisher, "Heliotherapy in the Treatment of Laryngeal Tuberculosis," 1067–68; B. B. Parish, "Tuberculosis of the Larynx," *United States Veterans' Bureau Medical Bulletin* 7 (1931): 238, Ryan, 73–75; Mayer, *Clinical Application*, 224–25; Victor X-Ray, *A Manual of Standardized Operative Technic for Users of Victor Ultraviolet Lamps*, 1929, General Electric Trade Literature, National Museum of American History, Smithsonian Institution. See also John R. Caulk and F. H. Ewerhardt, "Direct Internal Irradiation of Ultraviolet to the Bladder," *Archives of Physical Therapy, X-Ray, Radium* 13 (1932): 325–27; Oscar B. Nugent, "'Cold Quartz' Ultraviolet Orificial Irradiation," *Archives of Physical Therapy, X-Ray, Radium* 13 (1932): 82–86; Goldberg, "Heliotherapy," 283; S. L. Wang, "Quartz Light Therapy in Urogenital Tuberculosis," *JAMA* 104 (1935): 721.

69. Elizabeth Blaine Jenkins, "The Great Sky Medicine," *Hygeia* 5 (1927): 204.

70. Sara Sloane McCarty, "The Unfortunate Egglebert Ploot," *Hygeia* 14 (1936): 168.

Chapter Four

1. John Steele, "Doctors Again Giving King George New Treatment," *Chicago Daily Tribune*, 5 February 1929, 16; "Finds Warning for Chicago in King's Illness," *Chicago Daily Tribune*, 9 December 1928, 24; "King Continues Gain: Gloucester Home, Royal Family United," *New York Times*, 25 December 1928, 1; "King George Gains at Once at Seaside," *New York Times*, 11 February 1929, 6; "King Holds His Gain: Worst Phases Over, His Doctors Reveal," *New York Times*, 20 December 1928, 11; "King Rallies Again: New Drug to Be Used," *New York Times*, 29 December 1928, 1, 6; John Steele, "King Restless, Still in Pain, but Shows Gain," *Chicago Daily Tribune*, 17 December 1928, 1; "King Will Be Moved to a Seaside House," *New York Times*, 23 January 1929, 8; John Steele, "King's Fight on Illness Shows New Progress," *Chicago Daily Tribune*, 20 December 1928, 1.

2. See the following articles by W. A. Evans: "How to Keep Well: Sun Cure for T.B.," *Chicago Daily Tribune*, 26 March 1919, 8; "How to Keep Well: The Sun Cure," *Chicago Daily Tribune*,

17 December 1921, 8; "How to Keep Well: Vacation Sunlight," *Chicago Daily Tribune*, 23 August 1915, 6.

3. See W. A. Evans, "How to Keep Well: Health from Sunlight," *Chicago Daily Tribune*, 26 June 1924, 4, and his "How to Keep Well: Phenomena of Light," *Chicago Daily Tribune*, 9 June 1921, 8.

4. See the W. A. Evans pieces "How to Keep Well: Glass that Passes in Health Rays," *Chicago Daily Tribune*, 17 February 1928, 10; "How to Keep Well: Light Treatments," 8; "How to Keep Well: Skylight and Air Batting for the Sun," *Chicago Daily Tribune*, 1 January 1929, 14; "How to Keep Well: Sunlight," *Chicago Daily Tribune*, 13 October 1925, 8.

5. See the following articles by Lulu Hunt Peters: "Diet and Health: Sunlight and the Ultra Violet Rays," *Los Angeles Times*, 3 June 1927, A7; "Diet and Health: Infantile Paralysis," *Los Angeles Times*, 24 February 1927, A7; "Diet and Health: Running Ears," *Los Angeles Times*, 4 November 1926, A7; "Diet and Health: Dry Scaly Skin," *Los Angeles Times*, 18 October 1926, A6; "Diet and Health: T.B., Kidneys, and Bladder Trouble," *Los Angeles Times*, 30 September 1926, A7; "Diet and Health: Rickets and Sunshine," A6.

6. It is interesting to think about this move in light of Gail Bederman's *Manliness and Civilization: A Cultural History of Gender and Race in the United States, 1880–1917* (Chicago: University Press of Chicago, 1995), which argues that white people developed a theory in which they were the height of civilization. Their imperial success was evidence that they were fittest and deserved dominion over the planet. In essence, manliness and civilization came together under a single theory. In some respects, the sun celebrators at whom this chapter looks believed that primitive people were at the pinnacle of manly toughness. In other respects, however, Bederman's argument seems to hold.

7. Elizabeth M. Anderson, *The Essentials of Ultraviolet Treatment for the Use of Masseuses* (London: Balliere, Tindall and Cox, 1935), 1–2; Herman Goodman, *The Basis of Light in Therapy* (New York City: Medical Lay Press, 1926), 10–14; Hall, *Ultra-Violet Rays*, 1–3; Humphris, *Artificial Sunlight and Its Therapeutic Uses*, 3–5; Krusen, *Light Therapy*, 3–5; Russell and Russell, *Ultra-Violet Radiation and Actinotherapy*, 43–53.

8. Matthew B. Luckiesh, *Lighting the Home* (New York: Century Co., 1920), 3; Luckiesh, *Light and Work*, 7; see also, Luckiesh, *Artificial Sunlight*, chapter 1.

9. Aitken, *Ultra-Violet Radiations and Their Uses*, 2; Hall, *Ultra-Violet Rays*, 3; Rollier and Rosselet, *Heliotherapy*, v, 1–2.

10. Rollier and Rosselet, *Heliotherapy*, foreword; De Kruif, *Men against Death*, chapter 11; Merck, "Ancient and Modern Sun Worship," *Hygeia* 9 (1931): 602.

11. Stuart Chase, "Confessions of a Sun-Worshiper," *Nation*, 26 June 1929, 762–65.

12. Gladys Huntington Bevans, "Take Sun with Discretion," *Chicago Daily Tribune*, 20 July 1930, D3; "Sun Devotionals," *New York Times*, 6 July 1929, 12; Mildred Adams, "Modern Worshipers of That Old God, the Sun," *New York Times*, 7 July 1929, 70; "Sungods and Ultra-Violet Rays," *Electrical World*, 6 October 1928, 670; "School with Glass Walls Will Be Built in Berlin," *New York Times*, 30 October 1927, 1; C. I. Reed, "Old Sun-Worship Finds Vindication in Modern Laboratory," *Nation's Health* 8 (1926): 535–36.

13. Howland, "Starving for Sunshine," 16.

14. Edgar Lloyd Hampton, "A New Race of Sun Worshipers," *Los Angeles Times*, 2 January 1929, Annual Midwinter Supplement, I17; Ransome Sutton, "A Place in the Sun and What It Means," *Los Angeles Times*, 2 January 1930, C3.

15. "Germany's Quest of Greater Strength and Beauty," *Washington Post*, 13 December 1925, SM9; "Athletics Rising Greatly in Soviet [Union]," *New York Times*, 31 August 1930, E4; Anna Edinger, "Sunning and Airing for Health," *Hygeia* 1 (1923): 347–48; L. V. Dodds, "Sunlight in Industry," *Science Progress* 22 (1928): 655–59.

16. Thomas Quinn Beesley, "Smokeless New York and Smoggy London," *New York Times*, 23 October 1921, 42; Saleeby, *Eugenic Prospect*, 27–28. In *Crusaders for Fitness*, James Whorton talks a bit about the importance of hygiene in some evolutionary theories, but the thinking and time period he covers are considerably different from mine. The role of race in geographical thinking is not new. For a comprehensive, if a bit overwhelming, overview, see Frank A. Barrett, *Disease and Geography: The History of an Idea* (Toronto: Geographical Monographs, 2000). The essays in Nicholas A. Rupke, ed., *Medical Geography in Historical Perspective* (London: Wellcome Trust Center for the History of Medicine at UCL, 2000), are additional resources for transnational medical geographic thinking. See also, David N. Livingstone, "The Moral Discourse of Climate: Historical Consideration on Race, Place, and Virtue," *Journal of Historical Geography* 17 (1991), 413–34. For more about holistic geographical thinking, see Warwick Anderson, "Natural Histories of Infectious Disease: Ecological Vision in Twentieth-Century Biomedical Science," *Osiris* 19 (2004): 39–61; and for generalized anxieties about progress and place, see Charles Rosenberg, "Pathologies of Progress: The Idea of Civilization at Risk," *Bulletin of the History of Medicine* 72 (1998): 714–30.

17. Saleeby, *Eugenic Prospect*, 97–99.

18. See Saleeby's *Sunlight and Health*, 90–91, chapter 8, and his "From Heliotherapy to Heliohygiene," 513–14.

19. John Harvey Kellogg, "Address of Welcome" (paper presented at the Third Race Betterment Conference, Battle Creek, Michigan, 2–6 January 1928), 2.

20. I refer to the following papers presented at the Third Race Betterment Conference, Battle Creek, Michigan, 2–6 January 1928: Dr. Horace LoGrasso, "The Therapeutic Wonders of Sunlight," 632–39; C. C. Little, "Shall We Live Longer and Should We?," 12; Bunker, "Light and Life," 620; William T. Anderson, "The Health-Promoting Rays of Sunlight," 621–31.

21. Benjamin Goldberg, *Procedures in Tuberculosis Control for the Dispensary, Home and Sanatorium* (Philadelphia: F. A. Davis Co., 1933), chapter 1.

22. Luckiesh and Pacini, *Light and Health*, 30; F. G. Murray, "Pigmentation, Sunlight, and Nutritional Disease," *American Anthropologist*, 36 (1934): 438–40; Saleeby, *Sunlight and Health*, 8.

23. For new class implications, see chapter 5. Stanley Hoffland, "Features: Tan Is the Smart Shade this Summer, and Olive Oil Will Cure Sunburn," *Los Angeles Times*, 18 July 1926, B7; "Heliotherapy," *Spunk*, 22; Craig, "Leysin," 17; Irving S. Cutter, "How to Keep Well: Sunlight and Health," *Chicago Daily Tribune*, 26 June 1934, 10; Antoinette Donnelly, "Plenty of Sunshine Does Away with the Pale, Listless Look," *Chicago Daily Tribune*, 15 July 1931, 18; Gladys Huntington Bevans, "Sun Baths are Valuable if They Are Given Correctly," *Chicago Daily Tribune*, 31 July 1929, 31. For a look at changing senses of beauty, see Patricia Marks, *Bicycles, Bangs and Bloomers: The New Woman in the Popular Press* (Lexington: University Press of Kentucky, 1990), and Lois Banner, *American Beauty* (New York City: Alfred A. Knopf, 1983). Both authors write about the shifting sense of beauty, from pallid to glowing. Banner even addresses the growing popularity of tanning in the late nineteenth century and its eventual emergence as a widespread phenomenon after. Neither, however, provides the public health context that I offer here. Lears, *Fables of Abundance* (especially chapter 6) writes about the rising popularity of the tan and the boundaries

advertisers maintained (in their own work and in society more generally) between bronzed savages and white civilizers.

24. For a discussion of earlier thinking about climate and skin color, see Sheehan, "Doctor Photo."

25. See Donald C. Stockbarger, "The Health-Giving Ultraviolet Rays," 111–12, 116, and Wayne D. Heydecker, "Health and Housing," 125–33, both papers presented at the Tenth National Conference on Housing, Philadelphia, 28–30 January 1929.

26. Laurens, *Physiological Effects of Radiant Energy*, 11; "House Follows the Sun," *New York Times*, 4 June 1928, 14; Parsons, "Sun Worship," 213; Adams, "Modern Worshipers of That Old God, the Sun," 70; "Homes on the Roofs," *New York Times*, 24 February 1924, XX2.

27. James L. Holton, "News and Comment: Lighting the Airshafts," *Magazine of Light*, May 1935, 3; "Pipe-Lines for Sunshine," *Literary Digest*, 19 December 1931, 15–16.

28. Millar, Free, and Elway, *Sunrays and Health*, 98–99, "Homes on the Roofs," XX2.

29. See the building advertisement "500 Fifth Avenue Now Ready for Occupancy," *New York Times*, 7 January 1931, 23; also A. Lawrence Kocher and Gerhard Ziegler, "Sunlight Towers: Designs and Plans," *Architectural Record*, March 1930, 286–88; A. Lawrence Kocher and Gerhard Ziegler, "Sunlight Towers," *Architectural Record*, March 1929, 307–10; Parsons, "Sun Worship," 213; Henry D. Wright, *Rehousing Urban America* (New York: Columbia University Press, 1935), 135. Bradley's *The Works* treats developments in industrial design, including a section on natural lighting and sawtooth design.

30. Louise Bargelt, "Sparkling Glass Transforms a Breezy Porch to Radiant, Sun Filled Room for Year Round Use," *Chicago Daily Tribune*, 29 October 1933, C6; Philip M. Lovell, "The Care of the Body: Glass Houses," *Los Angeles Times*, 23 December 1928, I26; Philip Lovell, "Care of the Body," *Los Angeles Times*, 19 July 1925, L24; Louise Bargelt, "Home Building and Remodeling: Get Sun Tan in Your House Via 'Glass of Life,'" *Chicago Daily Tribune*, 4 August 1929, B2; F. Rose, "Prodigal Sun," *Delineator*, April 1925, 67; Century Apartments, "Summer of Winter," *New York Times*, 5 June 1932, RE4; Century Apartments, "2 Rooms to 7," *New York Times*, 14 June 1932, 18; Majestic Apartments, "Unique in Their Sheer Luxury," *New York Times*, 29 May 1932, RE5; London Terrace, "The Way to Live!," *New York Times*, 16 August 1932, 11; Butler and Baldwin, Inc., "The San Carlos," *New York Times*, 11 September 1927, RE11.

31. "Children Are Real Gainers by Suburban Home," *Chicago Daily Tribune*, 11 September 1927, L2; W. A. Evans, "How to Keep Well: Ultraviolet Ray Values," *Chicago Daily Tribune*, 7 August 1929, 14; Scarsdale Downs, "At Scarsdale Downs," *New York Times*, 26 July 1931, RE8; "Average House for Average Family Offers All Modern Essentials," *New York Times*, 21 July 1929, 142; Cord Meyer Development Company, "Live Smartly . . . Save money . . . Come to Forest Hills!," *New York Times*, 6 April 1930, 175; *Forest Hills Gardens*, 2nd ed. (New York: Sage Foundation Homes Co., 1913). In attempting to tie suburbia to health and equate challenges, moral and physical, with cities, Vita Glass drew on a long tradition in American history; see Kenneth T. Jackson, *Crabgrass Frontier: The Suburbanization of the United States* (New York: Oxford University Press, 1985); Gwendolyn Wright, *Building the Dream: A Social History of Housing in America* (Cambridge: MIT Press, 1998); Robert Fishman, *Bourgeois Utopias: The Rise and Fall of Suburbia* (New York: Basic Books, 1987).

32. See the following Vita Glass promotions: "Suburbs of the Future Build with Living Sunshine—for in Vita Glass a New Building Material Is Available to You Now," *New York Times*, 6 May 1929, 16; "Health Knocked . . . at Your Window Today . . . and You Refused Its Golden

Gift," *New York Times*, 22 April 1929, 20; "The Dynamo of Solar Energy Is Available for That Home You Will Build in the Fall," *New York Times*, 20 May 1929, 20; also Quartz-Lite, "—The Low Price," 11.

33. Ransome Sutton, "What's New in the Progress of Science: Will We Become Cave Dwellers?," *Los Angeles Times*, 29 April 1934, G14; Timothy G. Turner, "Our Cities Should Be Underground!," *Los Angeles Times*, 1 September 1935, G5, 14; "Now the Windowless Building with Its Own Climate," *New York Times*, 10 August 1930, XX4; "Windowless Building Held to Be Practical," *Los Angeles Times*, 28 December 1930, D3; "The Perfect Synthetic Life," *New York Times*, 7 October 1931, 20.

34. "Housing Group Asks Board of Experts on City Planning," *New York Times*, 11 April 1928, 1; "Says Skyscraper Is Public Benefit," *New York Times*, 16 December 1928, RE1; "Simulated Sunlight," *Housing* 21 (1932): 68–69. See also "Study to Get Sunlight in Home," *New York Times*, 27 September 1929, 20; "Calls Towers of Cuneo Type Peril to Health," *Chicago Daily Tribune*, 16 November 1929, 4.

35. "Wants City Homes Two Rooms Deep for Light and Air," *New York Times*, 10 June 1927, 1; Vita Glass, "Housing Expert Tells State Board Windows Should Admit Ultra-Violet Rays," *New York Times*, 28 June 1927, 52; "Simulated Sunlight," 68; "Planned Housing Fights Disease," By the People, For the People: Posters from the WPA, 1936–1943, Prints and Photographs Division, Library of Congress (available online as part of the Library of Congress American Memory collections, http://memory.loc.gov/).

36. Catherine Wurster Bauer, *Modern Housing* (Boston: Houghton Mifflin Co., 1934), cvi, 45, 57,150, see also part 1. For an account of Bauer's vision and the sad fate of her reform attempt, see Gail Radford, *Modern Housing for America: Policy Struggles in the New Deal* (Chicago: University of Chicago Press, 1996).

37. See Lewis Mumford's *Technics and Civilization* (San Diego: Harcourt Brace and Co., 1963) and his *The Culture of Cities* (New York: Harcourt, Brace and World, 1938), especially 5, 396.

38. Mumford, *Technics and Civilization*, chapter 5, section 9; Mumford, *Culture of Cities*, chapter 7, section 3.

39. Langdon Post, "Housing Questions," *New York Times*, 27 September 1935, 20; "Housing Project Shown in a Model," *New York Times*, 19 December 1935, 9.

40. Nicholas Dagen Bloom, *Public Housing That Worked: New York in the Twentieth Century* (Philadelphia: University of Pennsylvania Press, 2008), appendix A; "Cornerstone Laid for PWA Housing," *New York Times*, 15 October 1936, 29; "Vladeck Houses," *New York Times*, 24 November 1939, 22; "Huge Housing Unit in Redhook Opened," *New York Times*, 5 July 1939, 1, 2; *East River Houses: Public Housing in East Harlem* (New York: New York City Housing Authority, 1941); and from the New York City Housing Authority Archives, the following correspondence from folder 1, box 53E5, the LaGuardia and Wagner Archives, CUNY, La Guardia Community College (special thanks to Nicholas Bloom for pointing me in the direction of these materials): George D. Brown to Mr. Wright, 24 June 1938; Henry Wright to Alfred Rheinstein, 1 July 1938; Henry N. Wright and Bruno Funaro to Mr. Harrison, July 11, 1938.

41. United States Federal Housing Administration, *Property Standards: Requirements for Mortgage Insurance under Title II of the National Housing Act* (Washington, DC: Government Printing Office, 1934), 10. In subsequent years, circular number 2 outlined regional variations on housing restrictions. In *Crabgrass Frontier*, Ken Jackson points out some of the more disturbing ways that the FHA influenced the housing market. By proclaiming integrated or primarily minor-

ity neighborhoods dangerous investments, the FHA helped to encourage lending practices that favored white, suburban homeowners.

42. Federal Housing Administration, *The Fifth Annual Report of the Federal Housing Administration* (Washington, DC: Government Printing Office, 1939).

43. United States Federal Housing Administration, *Property Standards*, 3.

44. *Infant Care* (Washington, DC: United States Department of Labor Children's Bureau, 1933), 42–47; *Prenatal Care* (Washington, DC: United States Department of Labor Children's Bureau, 1931), 11–13: *The Child from One to Six* (Washington, DC: United States Department of Labor Children's Bureau, 1931), 6, 27–29.

45. *Sunlight for Babies* (Washington, DC: United States Department of Labor Children's Bureau, 1931); *Sunlight for Babies* (Washington, DC: United States Department of Labor Children's Bureau, 1926): "Sunlight for Babies," *American Physical Education Review* 30 (1925): 520–22.

46. From box 382, Central File, Records of the Children's Bureau: Grace Abbott to Mrs. Kathryn Van Aken Burns, 27 October 1930; W. T. McDerran to US Dept. of Labor, 19 July 1932; Children's Bureau to W. T. McDerran, 23 July 1932.

47. From box 382, Central File, Records of the Children's Bureau: A. N. Johnson to Ruth O'Brien, Bureau of Home Economics, 27 April 1931; Roy Brenholts to Children's Bureau, 6 December 1930; P. Wick to US Dept. of Labor, 25 April 1930; Frances C. Rothert to P. Wick, 25 April 1930; Frances C. Rothert to Mrs. A. E. Beckman, 4 October 1930; Chase D. Stephens to US Dept. of Labor, Children's Bureau, 14 March 1931; Isabel S. Fresen to Children's Bureau; Raymond H. Thompson to Children's Bureau, 14 February 1931; Blanche M. Haines, director of the Maternity and Infancy Division, to Raymond H. Thompson, 18 February 1931; Mrs. Harold Pack to "Madame," 28 February 1929; United States Children's Bureau to Mrs. Harold Pack, 14 March 1929; Mrs. O. E. Dolven to Children's Bureau, 13 November 1935; Ella Oppenheimer to Mrs. O. E. Dolven, 30 November 1935; for letters on lamps, Mrs. Donald Rosenlieb to Children's Bureau, 21 January 1935; D. E. Hayward to the Children's Bureau, 7 January 1932; Blanche M. Haines to D. E. Hayward, 22 January 1932.

48. *Sun Babies*, film recording, Records of the Children's Bureau.

49. "A Proposed Community Health Educational Program for the Prevention and Cure of Rickets and Malnutrition" and "Work of the Judson Health Center for the Year 1929," Judson Health Center, box 41, Judson Health Center folder, Community Service Society, CU RBML; "Are You as Well as You Can Be," box 5, Christodora House Records, CU RBML.

50. "Ten Years of Mulberry Health Center," box 60, folder 367-7, Community Service Society, CU RBML.

51. "Nutrition at Work in a Health Program," 1 May 1923, and "Yearly Progress Report from 1927–1928," box 60, folder 367-7, Community Service Society, CU RBML; "Yearly Report October 1, 1930–September 1931 Mulberry Health Center" and "Report of an Anti-Rachitic Program," November 1930, box 60, June 1929–June 1937 folder, Child Health Program, Community Service Society, CU RBML; Calendar, 1925, box 307, Community Service Society, CU RBML.

52. *Rickets Must Go!*, *Keep Baby's Legs Straight and Strong*, *In Mulberry District, Sunshine Made the Difference in the Growth of These Chicks*, box 307, Community Service Society, CU RBML.

53. Historians have told of the increasingly large role afforded science and medicine in determining proper child-rearing practices; see Rima D. Apple, *Mothers and Medicine: A Social History of Infant Feeding* (Madison: University of Wisconsin Press, 1987); Molly Ladd-Taylor, *Mother-Work: Women, Child Welfare, and the State, 1890–1930* (Urbana: University of Chicago

Press, 1994). Ladd-Taylor (who also writes about the birth and growth of the Children's Bureau) argues specifically that, around the turn of the century, women advocated publicly for proper child-rearing practices, using their authority as mothers to bolster claims. In time, however, women's primary responsibility became their own family's health, and they ceded the role of public advocate and childcare expert to doctors and child-welfare professionals. Meckel also tells of the attempt to train up better mothers between 1910 and 1930. See also Paul De Kruif, "Sunshine, Open Air and Those Awful Colds in the Head," *Ladies' Home Journal*, December 1928, 109; "Sun through the Open Kitchen Window Aid," *Washington Post*, 24 October 1926, F11; *Prenatal Care*, 11, 48; W. A. Evans, "How to Keep Well: Ultra-Violet Rays for Coming Mothers," *Washington Post*, 1 February 1927, 14.

54. Gladys Huntington Bevans, "Wanted by Babies: Lots of Sunshine, Real and Bottled," *Chicago Daily Tribune*, 8 January 1932, 20; "Sunning the Baby Out-of-Doors," *Washington Post*, 17 May 1931, MF12; "Uncle Sam's Home Hints: Sunlight Good for Baby," *Washington Post*, 12 October 1930, A5; "Uncle Sam's Home Hints: Sun Baths," *Washington Post*, 24 November 1932, 8; Josephine Hemenway Kenyon's "Baby's Place in the Sun," *Good Housekeeping*, April 1934, 95, and her "Babies In Summer Time," *Good Housekeeping*, August 1930, 124; S. J. Baker, "Eating Sunshine," *Ladies' Home Journal*, March 1930, 127, 130; Bundesen, "Sunshine and Health," 48, 51; Gladys Huntington Bevans, "Summer Improves Health of Baby if Mother Is Careful of Essentials," *Chicago Daily Tribune*, 25 June 1934, 15.

55. Ruth O'Brien, *Sun Suits for Children* (Washington, DC: United States Department of Agriculture Bureau of Home Economics, 1928); Clarice Louisba Scott, *Ensembles for Sunny Days* (Washington, DC: United States Department of Agriculture Bureau of Home Economics, 1930); "Two Sun Suits," *Washington Post*, 25 August 1929, SM15; Frank McCoy, "Health and Diet Advice: Sunbathing Suits for Children," *Los Angeles Times*, 15 August 1929, A6; "Sun Baths at Play Time," *Washington Post*, 4 May 1930, SM15.

56. Climax Sun Suit Company, "The Sunbath Suit," *Good Housekeeping*, June 1928, 239; Vanta Baby Garments, "Sunshine Health Insurance for Your Child," *Hygeia* 11 (1933): 756; Hecht Co., *Washington Post*, 3 June 1928, A1; Climax Sun Suit Company, "Sturdy Health Your Best Gift to Them," *Good Housekeeping*, June 1929, 246; Carter's Infant's Wear, "Now Carter's Sun Suits," *Good Housekeeping*, June 1929, 251.

57. I cite the following Palm Beach Suits ads: "An Added Ounce of Energy to Keep You on Your Toes," *New York Times*, 26 June 1926, 25; "Play it Safe . . . and You Can't Lose," *Chicago Daily Tribune*, 28 May 1931, 29; and ". . . There's More than One Way to Shut Out the Air," *Chicago Daily Tribune*, 2 July 1931, 32; also W. W. Coblentz, Ralph Stair, and Charles Warren Schoffstall, "Some Measurements of the Transmission of Ultra-Violet Radiation through Various Fabrics," *United States Bureau of Standards Journal of Research* 1 (1928): 105–24; Millar, Free, and Elway, *Sunrays and Health*, chapter 5; "Doctor Views Mens Clothing as Health Peril," *Chicago Daily Tribune*, 11 October 1930, 15; see in addition Parsons, "Sun Worship."

58. "Health and Diet Advice: Let the Blessed Sunshine In," *Los Angeles Times*, 15 August 1928, A8; "Says Scanty Attire Makes Women Hardy," *New York Times*, 18 November 1926, 5; "Sees Health in Thin Attire," *New York Times*, 13 June 1925, 5; "Advises Girls to Roll Their Hose and Grow Healthier," *Chicago Daily Tribune*, 24 July 1925, 14. For entertaining, popular histories of the beach and swimsuits, see Lena Lencek and Gideon Bosker's studies *The Beach: The History of Paradise on Earth* (New York City: Viking, 1998) and *Making Waves: The Story of the American Swimsuit* (San Francisco: Chronicle Books, 1989).

59. Sturmer, *The Modern Sun Cult*, 200–201.

60. "Thin Garb Foe of Tuberculosis Bundesen Says," *Chicago Daily Tribune*, 9 August 1927, 27; W. A. Evans, "How to Keep Well: Men Don't Know How to Dress," *Chicago Daily Tribune*, 23 October 1928, 12; "Health in Clothing," *Washington Post*, 25 September 1928, 6.

61. Certainly, the postwar break with the past, the cultural freedom of the 1920s, and the desire to forget the sorrows of the war were not as extreme as the standard portrayal of flappers, booze, and jazz indicates. Though historians have effectively debunked the caricature of the 1920s as unreflectively roaring, they have not fundamentally challenged the increased liberalism of the time. For this revisionist portrait, see Roderick Nash, *The Nervous Generation: American Thought, 1917-1930* (Chicago: Rand McNally, 1970), Dumenil, *The Modern Temper*; Paula S. Fass, *The Damned and the Beautiful: American Youth in the 1920's* (Oxford: Oxford University Press, 1977). For an account of new modes of dress, see Jenna Weissman Joselit, *A Perfect Fit: Clothes, Character, and the Promise of America* (New York: Metropolitan Books, 2001).

62. "What Do You Say When She Says, I Think I Want a Health Lamp?," *Electrical Record*, August 1929, 64-65.

63. "The Eternal Question—Clothes," *Washington Post*, 19 November 1926, 6; Frank McCoy, "Health and Diet Advice: Sunbathing," *Los Angeles Times*, 3 August 1931, A6; "Science Explains What's Wrong with Our Clothes," *Washington Post*, 2 May 1926, SM6. The scholarship on nudism is limited. One article that does a nice job synthesizing some of the reasoning behind early nudist philosophy is Ruth Barcan, "'Regaining What Mankind Has Lost through Civilisation': Early Nudism and Ambivalent Moderns," *Fashion Theory* 8 (2004): 63-82.

64. "Germany's Quest of Greater Strength and Beauty," SM9; H. A. Keller, *Back to Nature: The Story of This Nude World* (no publication information); Jan Gay, *On Going Naked* (Garden City, NY: Garden City Publishing Co., 1932).

65. "Our Appeal to the Race," *Sun Bathing Review*, Summer 1933, 3; "Nudism in Our Time," *Sun Bathing Review* 3, Spring 1935, 6-7.

66. "Three Things We Demand: The Magna Carta of Nudism," *Nudist*, January 1934, 6, republished regularly.

67. I refer to the following pieces from the *Nudist*: Ilsley Boone, "The Radiates," May 1933, 8; Ralph Kingman, "The How, When and Where of Sun-Bathing," June 1933, 15; J. Henry Hallberg, "How the Sun Builds Health and Beauty," September 1933, 23; A Congregational Minister, "Slowing Down Life's Temperature," June 1933, 25; "Balance," editorial comment, August 1933, 5; Ilsley Boone, "Clothes and Human Well-Being," June 1933, 7-8.

68. Maurice Parmelee, *Nudism in Modern Life: The New Gymnosophy* (New York: Alfred A. Knopf, 1931); William W. Newcomb, *The Story of Nudism* (New York: Greenberg, 1934); Ilsley B. Boone, *The ABC of Nudism: An Illustrated Handbook of the Movement in America, Its Practice and Philosophy* (New York: Sunshine Book Co., 1934); "Will Man be an Animal Failure," editorial comment, *Nudist*, October 1935, 3.

69. A Congregational Minister, "Slowing Down Life's Temperature," 25, "Modern Revolt," *Nudist*, editorial comment, September 1933, 5; Henry Knight Miller, "Joyous Freedom," *Nudist*, March 1934, 11; Keller, *Back to Nature*.

70. Parmelee, *Nudism in Modern Life*, 213; Boone, *ABC of Nudism*, 19.

71. Winston Willis, "Nudism and Normality," *Nudist*, September 1934, 8-9; "Will Man be an Animal Failure," 1; Art Eastman, *The Nude Deal* (South Bend, IN: Anchor Press, 1935), 30.

72. Parmelee, *Nudism in Modern Life* 60; "Balance," 5. Historians of 1920s American culture have written of the rise of Freudianism (see Nash, *Nervous Generation*, and Dumenil, *The Modern Temper*). Two texts, one old and the other far more recent, offer fuller treatments of Freud's pop-

ularity in the United States: Nathan G. Hale Jr., *The Rise and Crisis of Psychoanalysis in the United States* (New York: Oxford University Press, 1995), and Frederick J. Hoffman, *Freudianism and the Literary Mind* (Baton Rouge: Louisiana State University Press, 1967).

73. I. O. Evans, *Sensible Sun-Bathing* (London: Wyman and Sons, 1935).

74. Frances Merrill and Mason Merrill, *Nudism Comes to America* (New York: Alfred A. Knopf, 1932), 152, chapters 8–10, "Balance," 5.

75. Curtis was probably referring to the highly reclusive Tarahumara. Whether his visit was actual or invented is unclear, but the Tarahumara are legendary runners. According to stories of their past feats, they did hunt deer on foot and without weapons. Working in small groups, they chased the animals over an extended period, affording them no rest. After the long pursuit, the deer would become so tired that they could no longer run or fight back. Today, the Tarahumara are regarded as special for the small community of ultramarathoners (specialists in distances longer than 26.2 miles). Some admirers have even made pilgrimages like Curtis's to meet some of the world's least known but most remarkable athletes. See Scott Carrier, *Running after Antelope* (Washington, DC: Counterpoint, 2001), and Christopher McDougall, *Born to Run: A Hidden Tribe, Superathletes, and the Greatest Race the World Has Never Seen* (New York: Alfred A. Knopf, 2009).

76. Harry Ellington Curtis, "Lessons from the Tarahumares," *Nudist*, October 1934, 24; Boone, "The Radiates," 8.

77. Newcomb, *Story of Nudism*, 33–35; and the following listings from the *Nudist*: "Directory," June 1933, 29; "Our Nudist Directory," December 1933; "I.N.C. Nudist Directory," January 1935.

78. J. Henry Hallberg, "Signs along the Way," *Nudist*, September 1933, 20; "It's a Dark Day for Sun Bathing: Mayor Says 'No,'" *Chicago Daily Tribune*, 22 March 1932, 4; "Chicago Sun Bathers Must Wear Something," *New York Times*, 22 March 1932, 4; "Chicago's Nude Sun Bathers Arouse Ire," *Washington Post*, 21 March 1932, 10; "Nude Bathing Issue Shelved in Chicago," *Chicago Daily Tribune*, 22 March 1932, 2.

79. "The New Crusade," *Nation*, 16 January 1935, 62; "Nudists See Smith 'Inconsistent' Foe," *New York Times*, 4 January 1935, 26; "Catholics Deny Wide Reform Aim," *New York Times*, 15 January 1935, 23; Boone, *ABC of Nudism*, chapter 5; "Bay City Sets Nudist Roles," *Los Angeles Times*, 10 April 1934, 12; "Bad Nu(d)es for Nudists! Rules to Bare in Mind," *San Francisco Chronicle*, 10 April 1934, 4.

80. "Century of Progress Notes," *Chicago Daily Tribune*, 18 August 1934, 5; "Police at Fair Search Village, Seize 5 Nudists," *Chicago Daily Tribune*, 22 August 1934, 10; "Nudist Wedding Couple Is Jailed during Bail Hunt," *Chicago Daily Tribune*, 7 July 1934, 6; Percy Wood, "Nudist Couple 'Wed' in Jungle Uproar at Fair," *Chicago Daily Tribune*, 30 June 1934, 1; "San Diego Fair Crowds Prefer Midget Colony," *Chicago Daily Tribune*, 8 September 1935, F6. The movie *Zoro Nudist Colony* is available online from the San Diego Historical Society, http://www.sandiego history.org/ (accessed 26 September 2006).

81. America's Exposition, San Diego, California, box 91, and California Pacific International Exposition, box 78, Larry Zim World's Fair Collection, National Museum of American History Archives, Smithsonian Institution.

82. Robert Ernst, *Weakness Is a Crime: The Life of Bernarr Macfadden* (Syracuse: Syracuse University Press, 1991); Mark Adams, *Mr. America* (New York: HarperCollins, 2009).

83. Ibid.

84. As I've suggested, these *Physical Culture* articles are many. For a sampling, see Helen Anderson Storey, "A Sunshine Playground for Every Home, May 1929, 76; Frank Stilwell, "Sunshine Brings Radiant Health," May 1928, 46; Milo Hastings, "Sun-Tan All Winter Now," November 1930, 56; A. J. Lorenz, "The New Super-Food for Vitality," February 1929, 43; and three by Bernarr Macfadden himself: "Nude Cults Are Smashing Old-Time Prudery," April 1932, 32; "Clothes and Morals," January 1931, 44–45; and "Mankind Enslaved by Clothing," *Physical Culture,* June 1930, 30–31.

85. Bernarr Macfadden, *Macfadden's Encyclopedia of Physical Culture* (New York: Physical Culture Publishing Co., 1912), see especially 543–45, 1963–77, 2295–98, 2372–73; also two Macfadden articles in his *Encyclopedia of Health and Physical Culture* (New York: Macfadden Book Co., 1940): "Sunlight Treatment by Artificial Means," 2701–31, and "Sunlight a Foe to Disease," 2653–2700.

86. Parmelee, *Nudism in Modern Life,* 243.

Chapter Five

1. Leo McCarey, *The Awful Truth,* DVD (Sony Pictures, 2003).

2. For a solid history of vacationing in America, see Cyndy S. Aron, *Working at Play: A History of Vacations in the United States* (New York: Oxford University Press, 1999). Sheila Rothman and Hal Rothman have chosen different subjects, but their works each examine the role of rugged outdoor experience in travel and the sense that vacationing must be regenerative. See Sheila Rothman, *Living in the Shadow of Death,* and Hal Rothman, *Devil's Bargain: Tourism in the Twentieth-Century American West* (Lawrence: University Press of Kansas, 1998). Kevin Starr is the finest historian on California; see especially *Material Dreams: Southern California through the 1920s* (New York: Oxford University Press, 1990). For the history of health seekers in California specifically, see Emily Abel's *Suffering in the Land of Sunshine* and her *Tuberculosis and the Politics of Exclusion;* also John E. Baur, *The Health Seekers of Southern California, 1870–1900* (San Marino: Huntington Library, 1959).

3. An Old Promoter Forty Years in the Field of Real Estate, *Sunshine and Grief in Southern California: Where Good Men Go Wrong and Wise People Lose Their Money* (Detroit: St. Claire Publishing Co., 1931), 114. Contrary to the Old Promoter's contention, California tried to sell much more than just its sunny climate. Kevin Starr treats much of the state's self-promotion; see especially his *Inventing the Dream: California through the Progressive Era* (New York: Oxford University Press, 1985).

4. Dan Blanco and Erwin R. Schmidt, "Sunny Florida" (Chicago: Ted Browne Music Co., 1920); Miss Lola M. Emerson and A. W. Hall, "Tourist's Song of Florida" (St. Cloud: A. W. Hall, 1925), box 9, De Vincent Sheet Music Collection, National Museum of American History Archives, Smithsonian Institution; Ruth Cashen-Lippert, "Dedicated to the Sunshine City" (St. Petersburg: Mrs. Leroy P. Naylor, 1930), box 10, De Vincent Sheet Music Collection; Violet de Besa and Florence Howard-Millane, "Oh! California Sunshine!," (1925); Tom Ford and Mary Earl, "California Sunshine" (New York: Shapiro, Bernstein and Co., 1929); Geoffrey O-Hara, "The Sun Shines Bright in California" (Los Angeles: Wright Music Co., 1925); Harry Beloit, "Sunny California: You're Home Sweet Home to Me" (Los Angeles: Chas. F. Loveland, 1926); Fred Howard and Nat Vincent, "Make Your Mind Up to Wind Up in Sunny California" (Los Angeles: Morse M. Preeman, 1930); Francis V. Brady, "Golden California Sunshine" (Cleveland: Brady Music Co., 1920), box 3, De Vincent Sheet Music Collection.

5. An Old Promoter, *Sunshine and Grief*, 4–5; Clarita Dodge Corse, *Florida: Empire of the Sun* (Tallahassee: Florida State Hotel Commission, 1930), 7, 8, part 3.

6. "Sun's Actinic Rays are Magnet for Tourists," *New York Times*, 18 January 1925, XX18.

7. "Ultra Violet Rays Powerful in Miami," *Washington Post*, 7 December 1930, R5; "Sunshine Cure," *Wall Street Journal*, 14 December 1931, 16; "Where Winter Vacationists Dodge the Snow and Cold," *Chicago Daily Tribune*, 13 December 1931, 11; "Miami's Mayor Invites Chicago to Come South," *Chicago Daily Tribune*, 23 December 1934, E4; Carl Scheffel, "A Comparative Study of Ultraviolet Ray Intensity of Sun in Miami, Florida," *Medical Journal and Record* 136 (1932): 403–4; Putney Haight, "The Tribune Travelers' Guide: Hails Miami as America's One Great Cure-All," *Chicago Daily Tribune*, 3 February 1935, F5; "Miami Starts Second Summer Tourist Season," *Chicago Daily Tribune*, 7 June 1931, B2.

8. From the N. W. Ayer Advertising Agency Records, National Museum of American History Archives, Smithsonian Institution, see "Coral Gables" and "The Truth About Florida Real Estate," series 2, box 195; "Biting Blizzard or Balmy Breeze," "Announcing the Opening of the Coral Gables Office in This City," "Will You Take the Priceless Gift of Life?," and "To a Man Who Seeks the Opportunity of a Lifetime," series 2, box 194; "A Little Investment Grows Big and Strong in the Sunshine at Coral Gables," "Following the Sun-Track," and "Paths of the Sunwise," series 2, box 196.

9. "Sun-Ray Park Hotel Spa and Health Resort," Florida, box 3, Warshaw Collection of Business Americana, National Museum of American History Archives, Smithsonian Institution.

10. West Palm Beach, "Let Dr. Sun Protect Your Health at West Palm Beach," *New York Times*, 12 January 1930, 122; Miami, "Healthiest Spot In America Is Ready to Entertain You," *Chicago Daily Tribune*, 8 December 1929, I8; Miami, "Outwitting Winter in the Cities of the Sun," *New York Times*, 3 December 1933, XX5; Miami Beach, "Miami Beach on the Ocean in Florida's Sun!," *Chicago Daily Tribune*, 11 December 1927, D11; Miami Beach, "Give Them the Gift of Sun-Tanned Health!," *New York Times*, 15 January 1933, XX9; St. Petersburg, "Health in This Flood of Winter Sunshine," *Hygeia* 7 (1929): 1275.

11. Starr, *Material Dreams*; "Tourist Flow Maintained," *Los Angeles Times*, 5 January 1932, A2.

12. Clark Davis, "From Oasis to Metropolis: Southern California in a Changing Context of American Leisure," *Pacific Historical Review* 61 (1992): 357–86.

13. Edgar Lloyd Hampton, "Curses on Our Climate," *Los Angeles Times*, 26 January 1932, H8, 9; "Golden State-ments," *Los Angeles Times*, 16 June 1931, A4; R. E. Baugh, "Los Angeles, the Sunshine City," *Home Geographic Monthly*, June 1932, 25–30.

14. I refer to the following Los Angeles Times promotions: "'Rainbow Edition' of 6 Beautiful Magazines Will Be Out January 2nd," *Los Angeles Times*, 23 December 1928, 10; "Other New Features in the 'Rainbow Edition,'" *Los Angeles Times*, 28 December 1928, A11; "New Facts about Sunshine," *Los Angeles Times*, 20 December 1928, A7; "Send California's Sunshine to Your Friends Back East!," *Los Angeles Times*, 6 January 1929, B12; also Hampton, "A New Race of Sun Worshipers" and "An Economic Year Book," *Los Angeles Times*, 2 January 1929, Annual Midwinter Supplement, IV4.

15. San Francisco, "Make Winter Give You These Sunny Outdoor Days," *Good Housekeeping*, November 1931, 166; San Francisco, "Plan Now for This California Summer," *Good Housekeeping*, February 1932, 150; R. P. White, "Why the Desert Is a Magic Healer," *Los Angeles Times*, 18 November 1934, J4; "Death Valley's Floor Beckons," *Los Angeles Times*, 15 December 1933, A14.

16. William McGovern, "Spreading Tucson Sunshine," *Tucson*, December 1929, 7, 17; "Sunshine Making Tucson Famous," *Tucson*, March 1933, 9; Harry Hermann, "Elks State Hospital

Starts Works," *Tucson*, July 1932, 1–3, 12; "Health Palaces in Kingdom of Sunshine," *Tucson*, July 1931, 4, 5. For general historical background on Tucson, see Michael F. Logan, *Desert Cities: The Environmental History of Phoenix and Tucson* (Pittsburgh: University of Pittsburgh Press, 2006); C. L. Sonnichsen, *Tucson: The Life and Times of an American City* (Norman: University of Oklahoma Press, 1982).

17. Harold Bell Wright, "Why I Did Not Die," *American Magazine*, June 1924, 13.

18. For San Antonio, see "South with the Sun to San Antonio," *Good Housekeeping*, February 1930, 130, and "It's Summer All Winter at America's Most Picturesque Frontier!," *Chicago Daily Tribune*, 29 December 1929, F5. For Colorado: "Come and Play in the Land of Sunshine and Vitamins," *Chicago Daily Tribune*, 7 May 1930, 22, and "Whether You're Fun-Minded or Business-Minded Here's Your Finest Vacation," *Chicago Daily Tribune*, 23 April 1930, 21. For Michigan: "Magic Land for Children," *Chicago Daily Tribune*, 2 June 1929, H9; "There's No Other Summerland in the World Like Michigan," *Chicago Daily Tribune*, 20 April 1930, G11; and "Why Advertise West Michigan?," *Chicago Daily Tribune*, 15 June 1930, G14.

19. "You and the Sea and Vitamin D," "Sunshine Cruises," "Children of the Sun," series 3, box 235, N. W. Ayer Advertising Agency Records, Burlington Route; "Now . . . On 2 Famous Trains," *Chicago Daily Tribune*, 21 March 1928, 19.

20. Unguentine advertisements: "Stay Out in the Sun! Sunburn Vanquished Instantly," *Chicago Daily Tribune*, 19 August 1928, C9; "Tone Up Your Body in the Sun—but Beware Painful Sunburn," *Los Angeles Times*, 18 August 1929, I3; "*Get Out in the Summer Sun*, but—Avoid Needless Torture of Burning," *Los Angeles Times*, 21 July 1929, 16; "Get a *Healthy* Tan with None of Sunburn's Torture," *Los Angeles Times*, 30 June 1929, I6.

21. Lucky Strike advertisements: "Sunshine Mellows, Heat Purifies," *Chicago Daily Tribune*, 28 February 1931, 8; "Sunshine Mellows, Heat Purifies," *Los Angeles Times*, 13 January 1931, A8; "Sunshine Mellows, Heat Purifies," *Washington Post*, 10 March 1931, 4; "'A Significant Example,'" *New York Times*, 18 September 1930, 30; "'Making Friends and Holding Them,'" *Los Angeles Times*, 16 September 1930, 8; "'Making Friends and Holding Them,'" *Washington Post*, 16 September 1930, 4.

22. *What Everyone Should Know about Ultra-Violet Rays*, Hanovia Chemical and Manufacturing Company, Trade Literature, National Museum of American History, Smithsonian Institution; Vita Glass, "Where Did you Get That Tan . . . ?," *New York Times*, 11 March 1929, 17, and "Where Did You Get That Tan . . . ?," *Wall Street Journal*, 19 March 1929, 6; General Electric, "You Can Have a Summer Seaside Tan," X15; Commonwealth Edison Electric Shops, "Florida Sunshine at Home," 6, and "Bring Florida Sunshine to Chicago for the Winter Months," *Chicago Daily Tribune*, 28 January 1930, 40.

23. Pfund, "A Practical Window for Transmitting Ultraviolet Rays," 18–19.

24. In *Face Value: The Politics of Beauty* (Boston, Routledge and Kegan Paul, 1984), Robin Tolmach Lakoff and Raquel L. Scherr most fully argue this point, but their reading of the evidence is, in many respects, problematic. They look at how beauty and gender intersect and conclude that the tanned woman that emerges as an aesthetic ideal was beautiful, in large part, because she was idle: "The tan now indicates higher social, position, and therefore health (despite its link to skin cancer) and beauty (despite its link to premature wrinkles)," 177–78. While glamorous celebrities support large parts of this argument, and indeed the tan was a marker of time spent relaxing, it was not always about idleness and it was not solely for women. For men, time in the sun was time recharging for work, and for women, time in the sun was often associated with vigorous and active bodies. These are far different objectives from those that Scherr and

Lakoff identify and are deeply problematic for their conclusion that women, with their aesthetic choices, try to distinguish themselves from men. The tanned, sporting woman was beginning to join activities that had once been decidedly male. My evidence calls some of their other conclusions into doubt too. When tan became beautiful, it was far more a sign of good health than an indicator of potentially cancerous mutations. Americans of all classes were supposed to find their way to a bronzed complexion. Finally, in large part because it was healthful, tan was an aesthetic associated with vigor not idle leisure (for women and, even more, for men).

25. See the Southern California promotions "Make Your Winter Vacation Benefit Your Business," *Wall Street Journal*, 10 December 1930, 7, and "A Word to the Wives Whose Husbands Are Worried about Business," *Good Housekeeping*, December 1932, 169; Hanovia's "You Need Vigor for Success," *New York Times*, 24 November 1935, E10; and the following Crystal Health Club advertisements: "What *Every* 'He-Man' Needs!," *New York Times*, 10 December 1935, 47; "March of Time Pushes Executives Out of Line," *New York Times*, 12 January 1934, 43; "Danger Lurks in Every Executive's Life," *New York Times*, 9 October 1934, 36; "20,000 Alert Executives Can't Be Wrong," *New York Times*, 5 January 1934, 30.

26. I refer to the following Parke-Davis promotions: "Sunlight That Comes up from the Sea in Ships," *Good Housekeeping*, December 1929, 107; "Golden Gallons of Sunshine from a Wintry Sea," *Good Housekeeping*, March 1930, 105; "Vitamin Farms in the Icy Seas," *Good Housekeeping*, November 1929, 105. See also Murray, "Pigmentation, Sunlight, and Nutritional Disease," 441, Krusen, *Light Therapy*, 56–57.

27. "A Secret Kept for Centuries," series 2, box 98, N. W. Ayer Advertising Agency Records; and the following Squibb promotions: "Soft Bones, Poor Teeth, Now Declared Common," *Good Housekeeping*, January 1927, 144; "Your Baby Must Have Bottled Sunshine," *Good Housekeeping*, November 1927, 200; "Your Baby Must Have Bottled Sunshine," *Chicago Daily Tribune*, 16 October 1927, J6; "Your Baby May Look Strong . . . Plump . . . Rosy," *New York Times*, 13 November 1927, RP6; "He's Building His Bones and Teeth Now!," *Chicago Daily Tribune*, 5 February 1928, C7; "For Strong, Solid Little Bodies . . . Rosy Cheeks," *Chicago Daily Tribune*, 11 October 1928, 17; "Babies . . . and Bottled Sunshine," *Good Housekeeping*, January 1929, 183; "Babies . . . and Bottled Sunshine," *Los Angeles Times*, 6 January 1929, J5; "The First Two Years Give the Answer," *Good Housekeeping*, December 1929, 229; "Few Babies Get Enough Sunlight," *Good Housekeeping*, March 1927, 114. See also Marchand, *Advertising the American Dream*, chapter 7.

28. "He Shall Not Go Naked into Battle," "The Light That Shines for All," "They Gave Man Forty Added Years," series 11, box 41, N. W. Ayer Advertising Agency Records.

29. From the N. W. Ayer Advertising Agency Records, see "A Source of Strength, a Wall of Defense—*for the Growing Child, the Invalid, the Convalescent*," "Children and Adults Too—," "Help Him Resist Infection and Disease," and "The Richest Available Source of Health-Building Vitamins!," series 2, box 97; "A Friendly Substitute for Sunshine," series 2, box 316; and "A Nameless Puppy Helped Snare the Sun," series 11, box 41.

30. I refer to the Parke-Davis advertisement "Spoonfuls of Summer Sun . . . for Winter's Cold, Dark Days," *Hygeia* 8 (1930): 6; also the following Squibb promotions: "How Many of These Ten," *Chicago Daily Tribune*, 8 January 1928, C6: "Babies . . . and Bottled Sunshine," *Good Housekeeping*, 183; "Babies . . . and Bottled Sunshine," *Los Angeles Times*, J5; "Bottled Sunshine! A Necessity—*Say Authorities—to Build for Your Baby*," *Good Housekeeping*, October 1927, 151.

31. Parke-Davis, "Vitamin Farms in the Icy Seas," 105; Squibb, "These for Your Baby with the Help of Bottled Sunshine," *Good Housekeeping*, October 1931, 126; Coco Cod, "Tastes Like

Chocolate," *Chicago Daily Tribune*, 4 April 1926, J4; McCoy's Cod Liver Tablets, "Oh Mamma! I Can't Take Cod Liver Oil," *Good Housekeeping*, January 1931, 127, and "Don't Force Child to Swallow Oil," *Good Housekeeping*, February 1931, 250; White's Cod Liver Oil Concentrate, "Cod Liver Oil Nightmare," *Good Housekeeping*, December 1933, 160.

32. Alfred F. Hess and Mildred Weinstock, "Antirachitic Properties Imparted to Inert Fluids and to Green Vegetables by Ultra-Violet Radiation," *Journal of Biological Chemistry* 62 (1924): 301–13; H. Steenbock and A. Black, "Fat-Soluble Vitamins," part 17, "The Introduction of Growth Promoting and Calcifying Properties in a Ration by Exposure to Ultra-Violet Light," *Journal of Biological Chemistry* 61 (1924): 405–22. See also Alfred F. Hess, Mildred Weinstock, and F. Dorothy Helman, "The Antirachitic Value of Irradiated Phytosterol and Cholesterol," part 1, *Journal of Biological Chemistry* 63 (1925): 305–8.

33. A. D. Emmett, "Laboratory Notes: Vitamin D, 1923–1931," 513–14, box 84, Parke-Davis Research Laboratory Records, National Museum of American History Archives, Smithsonian Institution.

34. Ibid.

35. Norman E. Bowie, *University-Business Partnerships: An Assessment* (Lanham, MD: Rowman and Littlefield, 1994), 5–7; David Blumenthal, Sherrie Epstein, and James Maxwell, "Commercializing University Research: Lessons from the Experience of the Wisconsin Alumni Research Foundation," *New England Journal of Medicine* 314 (1986): 1621–23; Rima D. Apple's *Vitamania*, 51–58, and her "Patenting University Research: Harry Steenbock and the Wisconsin Alumni Research Foundation," *Isis* 80 (1989): 374–94.

36. I refer to the following articles by W. A. Evans: "How to Keep Well: Sunlight," *Washington Post*, 14; "How to Keep Well: Sunlight," *Chicago Daily Tribune*, 8; and "How to Keep Well: Explorations in Light," *Chicago Daily Tribune*, 8 May 1925, 8. See also Edward T. Wilkes, Daye W. Follett, and Eleanor Marples, "The Treatment of Rickets with Irradiated Ergosterol," *American Journal of the Diseases of Children* 37 (1929): 483–96; "Ultra-Violet Light Used to Enrich Food," *New York Times*, 6 July 1930, 14; Baker, "Eating Sunshine," 127–30; Earl W. May, "The Prevention of Rickets in Premature Infants by the Use of Viosterol 100 D," *JAMA* 96 (1931): 1376–80.

37. Emmett, "Laboratory Notes," 458, 460, 512.

38. "Committee Work Vitamins A & D U.S.P. & Miscellaneous Committees, 1921–1938," 69–77, 82–124, 416–18, 422–23, box 84, Parke-Davis Research Laboratory Records.

39. Ransome Sutton, "YOU May Be Starving for This Tiny Crystal," *Los Angeles Times*, 16 June 1933, H4; Iva M. McFadden, "Let 'Sol' Do It," *Los Angeles Times*, 12 April 1931, J5; F. Rose, "Food That Imprisons Sunlight," *Delineator*, July 1926, 55.

40. Del Monte, *Sunshine Gatherers*, Motion Picture, Broadcasting, and Recorded Sound Division, Library of Congress.

41. For similar arguments, see Lears, *Fables of Abundance*, chapter 6.

42. Fleischmann's promotions: "Now in a Familiar Food . . . the Health Value of Hours in the Summer Sun," *Good Housekeeping*, September 1929, 120–21; "'It Is Natural for Me . . . a Skin Specialist . . . to Be an Advocate of Yeast,'" *Los Angeles Times*, 24 November 1929, I11; "'It Is Natural for Me . . . a Skin Specialist . . . to Be an Advocate of Yeast,'" *Chicago Daily Tribune*, 24 November 1929, E4; "For 'Sunlight-Starved' Men, Women and Children," *Los Angeles Times*, 22 September 1929, I5; "For 'Sunlight-Starved' Men, Women and Children," *Chicago Daily Tribune*, 22 September 1929, D4; "People Who Work Indoors Need This Health Protection," *Good Housekeeping*, October 1929, 140.

43. Fleischmann's promotions: "Now in a Familiar Food" 120–21; "Warns of Danger in Pace of Modern Living," *Chicago Daily Tribune*, 3 November 1929, F2; "Warns of Danger in Pace of Modern Living," *Los Angeles Times*, 3 November 1929, I11, "People Who Work Indoors," 140.

44. Katharine Blunt and Ruth Cowan, "Irradiated Foods and Irradiated Ergosterol," *JAMA* 93 (1929): 1301–8; Quaker Farina, "No Other Children's Cereal Brings a Full Hour of Sunlight's Health in Every Dish," *Good Housekeeping*, March 1930, 146, and "Known for Years as the First Solid Food for Infants," *Good Housekeeping*, January 1930, 200; Muffets, "Muffets . . . How They Crunch," *Chicago Daily Tribune*, 16 February 1929, 7.

45. Bond Bread promotions: "'Mothers, *Attention!*'" *Good Housekeeping*, May 1933, 153; "Grown-Ups, *Too!* Need Sunshine Vitamin-D," *Good Housekeeping*, April 1932, 111; "Scientists Agree . . . All Ages Need Sunshine Vitamin D," *Good Housekeeping*, February 1932, 108; "Sunshine Vitamin-D *Helps Build* Strong Bones," *Good Housekeeping*, March 1932, 92; "To the MOTHERS of Growing Children," *Good Housekeeping*, January 1932, 95.

46. Bond Bread promotions: "To the MOTHERS of Growing Children," 95; "Bond Bread is the *Only Food* that Gives You the Extra *Sunshine Vitamin-D* You Need," *New York Times*, 5 May 1931, 14; "Bond Bread Now Brings You the One Scarce Vitamin," *Washington Post*, 20 October 1931, 12. See also Emmett, "Laboratory Notes," 646.

47. Borden's, "How Much Sunshine Do You Buy?," *New York Times*, 14 December 1927, 34; Blue Valley Butter promotions: "'Are You Darkening Their Lives,'" *Chicago Daily Tribune*, 25 September 1928, 16; "Life's Race Demands 'Edible Sunshine,'" *Chicago Daily Tribune*, 19 May 1928, 9; "Nature Places in *Good* Butter the Concentrated Sunshine—Which City Children *Need for Growth and Health*," *Chicago Daily Tribune*, 10 March 1928, 9.

48. G. C. Supplee, "Antirachitic Activation of Milk by Direct Irradiation with Ultra-Violet Rays," *American Journal of Public Health* 23 (1933): 225–29; James A. Tobey, "Vitamin D Milk," *Hygeia* 11 (1933): 694–96; "Chicago Dairymen Plan 'Sunshine' Milk Supply," *New York Times*, 25 March 1934, E7; "Council Amends Law to Permit 'Sunshine Milk,'" *Chicago Daily Tribune*, 3 April 1934, 14; "Council Health Group Indorses 'Sunshine' Milk," *Chicago Daily Tribune*, 5 April 1934, 13; "Vitamin D Milk Approved," *New York Times*, 15 March 1935, 44; Borden's, "Borden's Announces Irradiated Vitamin D Milk," *Chicago Daily Tribune*, 22 April 1934, 20.

49. Borden's, "Now Available throughout Chicago," *Chicago Daily Tribune*, 25 April 1934, 13; Bowman Dairy Company, "Bowman's Vitamin D Milk Prevents Rickets, Helps Build Sound Teeth," *Chicago Daily Tribune*, 22 April 1934, 27; Pet Milk, "But the Sun Doesn't Always Shine," *Good Housekeeping*, August 1934, 125, and "Priceless for Her Future," *Good Housekeeping*, October 1935, 171; Wieland's, "Wieland's Announces Vitamin D Milk," *Chicago Daily Tribune*, 22 April 1934, 24. See also from the N. W. Ayer Advertising Agency Records "Milk Is *Not* a Cure-All, but It's the Greatest Protective Food Yet Discovered," series 3, oversize 347, and "Selected Milk from Sealtest," series 3, oversize 349

50. "Another Contribution to the Health of the Community," "Milk Is *Not* a Cure-All," "For Child Health, Follow the Advice of Uncle Sam," "Every Day is *Sunday* for Baby," and "The Benefits of Sunshine Vitamin for Your Family," series 3, oversize 347, N. W. Ayer Advertising Agency Records.

51. From the Products Cookbooks Collection, National Museum of American History Archives, Smithsonian Institution, see "From Soup to Dessert with the New Irradiated Pet Milk," box 11; "Summer Time Breakfasts, Lunches, Dinners Made with Irradiated Pet Milk," box 26; "Tempting Treats for Summer, for 2 or 4 or 6," box 27.

52. Milk box 1 image from *The Adventures of the Vita-Men: A-B-E and G with Their Leader Vit-Man D*, Warshaw Collection of Business Americana, National Museum of American History Archives, Smithsonian Institution; Carnation, "The Improved Carnation Milk with the 'Sunshine Vitamin,'" *Good Housekeeping*, July 1934, 175.

53. *Borden's Presents the Sunshine Makers*, Motion Picture, Broadcasting, and Recorded Sound Division, Library of Congress.

54. "Smokies" is the name I give to them, but it seems fitting.

55. *Borden's Presents the Sunshine Makers*.

56. "Advertising News and Notes: Schlitz Campaign Starts Today," *New York Times*, 22 May 1936, 32; Schlitz promotions: "For Refreshing Energy," *Chicago Daily Tribune*, 19 June 1936, 29; "Extra Refreshment in Schlitz," *Washington Post*, 31 July 1936, 20; "Extra Refreshment in Schlitz," *New York Times*, 10 July 1936, 20; "Announcing . . . Schlitz the Beer with Sunshine VITAMIN-D"; *New York Times*, 22 May 1936, 12; "*More* than Refreshing," *New York Times*, 19 June 1936, 14; "*More* than Refreshing," *Chicago Daily Tribune*, 7 August 1936, 8.

57. Pet Milk, "Here You Are! New and Better Milk—at No Extra Cost!," *Los Angeles Times*, 18 June 1934, 4; Carnation, "The Improved Carnation Milk with the 'Sunshine Vitamin,'" 175; Pet Milk, "Sunshine in Their Milk at No Increase in Cost!," *Chicago Daily Tribune*, 3 July 1934, 14.

58. At least it did not claim to be fortified, and the ad dating from 1932 probably would have featured D fortification if it could have.

59. I refer to the following promotions from the N. W. Ayer Advertising Agency Records: "Keep Your Vacation Vitality!," series 3, oversize 347; "The First Step to Summer Health All Winter" and "Summer Sun All Winter," series 3, oversize 433; "How to Keep That Vacation Vitality," series 3, oversize 346.

60. *Substitutes for the Sun* (Washington, DC: United States Department of Labor Children's Bureau, 1940).

61. "This Is the Way I like Vitamin D," "Hey Mom—I Think I'm Getting a Tooth," "I Asked My Doctor Why Vitamin D?," "What Are Little Boys Made Of?," "Pure Gold for Brown Bodies, " and "Who Says Winter's a Barrier? Extra Vitamin D Helps Him through the Cold Shut-In Days," series 3, oversize 434, N. W. Ayer Advertising Agency Records.

Epilogue

1. US Department of Health and Human Services, *Healthy People 2010: Understanding and Improving Health*, 2nd ed. (Washington, DC: US Government Printing Office, 2000), chapter 3. The section is poorly written, as it implies that getting one out of every four Americans to avoid tanning beds will be a success. Presumably, it hopes that at least 75 percent of Americans will use one of the other three preventive measures. See Melody J. Eide and Martin A. Weinstock, "Epidemiology of Skin Cancer," in *Cancer of the Skin*, ed. Darrell S. Rigel et al. (Philadelphia: Elsevier Saunders, 2005), 47–60.

2. Mark Dickie and Shelby Gerking, "The Formation of Risk Beliefs, Joint Production and Willingness to Pay to Avoid Skin Cancer," *Review of Economics and Statistics*, 78 (1996): 451–63.

3. "How to Get a Good Tan without the Sun," *Consumer Reports*, October 2004, 7.

4. Ed Giovannucci, interview by Ira Flatow, 17 June 2005, *Science Friday* (NPR).

5. M. Nathaniel Mead, "Benefits of Sunlight: A Bright Spot for Human Health," *Environmental Health Perspectives* 116 (2008): A160–67; Janet Raloff's "The Antibiotic Vitamin," *Science*

News 170 (2006): 312, 317, and her "Vitamin D: What's Enough?," *Science News* 166 (2004): 248–49; Erik Stokstad, "The Vitamin D Deficit," *Science*, n.s. 302 (2003): 1886–88; Tara Parker Pope, "The Miracle of Vitamin D: Sound Science or Hype?," *Well* (blog), *New York Times*, 1 February 2010, http://well.blogs.nytimes.com/2010/02/01/the-miracle-of-vitamin-d-sound -science-or-hype (accessed 19 February 2010).

6. "Seasonal Pattern Specifier," *DSM-IV* (Washington, DC: American Psychiatric Association, 1994), 389.

7. See the Parans website, http://www.parans.com/lighthealth.htm (accessed 23 January 2007).

8. Timothy Williams, "Here Comes the Sun, Redirected," *New York Times*, 2 June 2005, B2, corrected 14 June 2005.

9. Carpenter Norris Consulting, www.carpenternorris.com (accessed 18 December 2006); "Solar Light Pipe in Washington, D.C.," *Detail* 4 (2004), www.detail.de/rw_5_Archive_En _HoleArtikel_5331_Artikel.htm (accessed 18 December 2006); Alex Frangos, "Here Comes the Sun—Energy-Code Changes Spur Move to More Natural Light," *Wall Street Journal*, 12 November 2003, B1.

10. Michael Dumiak, "Lifting the Winter Dark: Mirrors to Reflect Light Into a Town That Gets No Sun," *Scientific American*, 3 April 2006, http://www.scientificamerican.com/article .cfm?id=lifting-the-winter-dark (accessed 9 August 2010).

INDEX

Italicized page numbers refer to figures.